Biology for the Logic Stage

Student Guide

Biology for the Logic Stage Student Guide

Second Edition (3rd Printing 2020)
Copyright @ Elemental Science, Inc.
Email: support@elementalscience.com

ISBN# 978-1-935614-39-5

Printed in the USA for worldwide distribution
Pictures by Paige Hudson and Erin Simons (One Line Design)

For more copies write to:
Elemental Science
PO Box 79
Niceville, FL 32588
support@elementalscience.com

Copyright Policy

Biology for the Logic Stage Student Guide
Table of Contents

Biology for the Logic Stage
Letter to the Student

Dear Student,

Welcome to your journey through biology, which is the study of living things. This year you will examine the various structures of life and how they operate. You will look at plants, animals, and the human body along your voyage. This guide is written to you, so enjoy your journey!

What does this guide contain?

First, this guide includes the Date Sheets and Unit Materials for each of the units. The Unit Materials include the Vocabulary Sheet for the unit, weekly Student Assignment Sheets, blank sketches, experiment sheets, and space for each of your writing assignments. After the Unit Materials, you will find the Appendix of this guide. In it you will find a list of all the memory work for the year, a glossary, and a place to record any additional activities you have done that pertain to biology.

Student Assignment Sheets

The Student Assignment Sheets contain your weekly assignments for each week. Each of the student assignment sheets contains the following:

✓ **Experiment** – Each week will revolve around a weekly topic. You will be assigned an experiment to complete that poses a question about what you are studying. Each Student Assignment Sheet contains the list of materials you will need and the instructions to complete the experiment. This guide also includes experiment sheets for you to fill out each week. In each of these experiments, you will use the scientific method.

   A Word about the Scientific Method —The scientific method is a method for asking and answering scientific questions. This is done through observation and experimentation. The following steps are key to the scientific method:

   1. **Ask a Question** – The scientific method begins with asking a question about something you observe. Your questions must be about something you can measure. Good questions begin with how, what, when, who, which, why, or where.
   2. **Do Some Research** – You need to read about the topic from your question so that you can have background knowledge of the topic. This will keep you from repeating the mistakes that have been made in the past.
   3. **Formulate a Hypothesis** – A hypothesis is an educated guess about the answer to your question. Your hypothesis must be easy to measure and answer the original question you asked.
   4. **Test with Experimentation** – Your experiment tests whether your hypothesis is true or false. It is important for your test to be fair. This means that you may need to run multiple tests. If you do, be sure to only change one factor at a time so that you can determine which factor is causing the difference.

Biology for the Logic Stage Student Guide ~ Introduction

5. **Record and Analyze Observations or Results** – Once your experiment is complete, you will collect and measure all your data to see if your hypothesis is true or false. Scientists often find that their hypothesis was false. If this is the case, they will formulate a new hypothesis and begin the process again until they are able to answer their question.

6. **Draw a Conclusion** – Once you have analyzed your results, you can make a statement about them. This statement communicates your results to others.

Each of your experiment sheets will begin with a question and an introduction. The introduction will give you some background knowledge for the experiment. The experiment sheet also contains sections for the materials, a hypothesis, a procedure, an observation, and a conclusion. In the materials section, you need to fill out what you used to complete the experiment. In the hypothesis section, you need to predict the answer to the question posed in the lab. In the procedure section, you need to write a step-by-step account of what you did during your experiment. In other words, you need to provide enough detial so that someone else could read your report and replicate your experiment. In the observation section, you need to write what you saw and observed as well as any results you measured. Finally in the conclusion section, you need to write whether or not your hypothesis was correct and any additional information you have learned from the experiment. If your hypothesis was not correct, discuss why with your teacher and then include why your experiment did not work on your experiment sheet.

Safety Advisory—Do not perform any of the experiments marked " ☻ **CAUTION** " on your own. Be sure you have adult supervision.

☐ **Vocabulary and Memory Work** – Throughout the year, you will be assigned vocabulary and memory work for each unit. Each week, you will need to look up the word in the glossary on pp. 255-262 and fill out the definitions on the Unit Vocabulary Sheet found at the beginning of each unit. You may also want to make flash cards to help you work on memorizing these words. Each week, you will also have a memory work selection. Simply repeat this selection until you have it memorized, and then say the selection to your teacher. There is a complete listing of the memory work selections in the Appendix on pp. 247-249.

▨ **Sketch** – Each week, you will be assigned a sketch to complete. Color the sketch and label it with the information given on the Student Assignment Sheet. Be sure to give your sketch a title.

✍ **Writing** – Each week, you will be writing an outline and/or a narrative summary. The student assignment page will give you a reading assignment for the topic from your spine text, either *The Usborne Science Encyclopedia* or *The Kingfisher Science Encyclopedia*. After you have finished the assignment, discuss what you have read with your teacher. Your teacher will let you know whether to write an outline or a narrative summary from your reading. Your teacher may also assign additional research reading out of the following books:

📖 *The Usborne Science Encyclopedia, 2015 Edition (USE)*
📖 *The Kingfisher Science Encyclopedia, 2017 Edition (KSE)*
📖 *DK Encyclopedia of Nature, 2007 Edition (DKEN)*
📖 *Usborne Illustrated Dictionary of Science, 2012 Edition (UIDS)*
(**Note**—*The editions noted here are the most current editions. However, the past two editions of each of these spines will also work.*)

Once you finish the additional reading, prepare a narrative summary about what you have learned from your reading. Your outlines should be one-level main topic style outlines and your narrative summaries should be one to three paragraphs in length, unless otherwise assigned by your teacher.

🕒 **Dates** —Each week, dates of important discoveries within the topic and dates from the readings are given on the student assignment sheet. You will enter these dates onto one of four date sheets. The date sheets are divided into the four time periods laid out in *The Well-Trained Mind* by Susan Wise Bauer and Jessie Wise (Ancients, Medieval-Early Renaissance, Late Renaissance-Early Modern, and Modern). These sheets are found in the ongoing projects section of this guide. You can choose to just write the dates and information on the sheet or you can draw a timeline in the space provided and enter your dates on that.

How to schedule this study

Biology for the Logic Stage is designed to take up to three hours per week. You, along with your teacher, can choose whether to complete the work over five days or over two days. Below are two options for scheduling to give you an idea of how you can schedule your week:

✓ A typical two-days-a-week schedule
  - 🕒 **Day 1** – Define the vocabulary, do the experiment, complete the experiment page, and record the dates.
  - 🕒 **Day 2** – Read assigned pages and discuss together, prepare the science report or outline, and complete the sketch.

✓ A typical five-days-a-week schedule
  - 🕒 **Day 1** – Do the experiment and complete the experiment page.
  - 🕒 **Day 2** – Record the dates and define the vocabulary.
  - 🕒 **Day 3** – Read assigned pages and discuss together and complete the sketch.
  - 🕒 **Day 4** – Prepare the science report or outline.
  - 🕒 **Day 5** – Complete one of the Want More activities from the Teacher Guide.

Additional Resources

The following page contains quick links to the activities suggested in this guide along with several helpful downloads:

🌐 https://elementalscience.com/blogs/resources/bls

Final Thoughts

As the author and publisher of this curriculum, I encourage you to contact me with any questions or problems that you might have concerning *Biology for the Logic Stage* at support@elementalscience.com. I will be more than happy to answer them as soon as I am able. I hope that you will enjoy *Biology for the Logic Stage*!

Sincerely,
Paige Hudson
BS Biochemistry, Author

Ancients 5000 BC-400 AD

Medieval-Early Renaissance 400AD-1600AD

Late Renaissance-Early Modern 1600 AD-1850 AD

Modern 1850 AD-Present

Biology Unit 1

Biological Building Blocks

Unit 1: Introduction & Ecology
Vocabulary Sheet

Define the following terms as they are assigned on the Student Assignment Sheet.

1. Cell – basic unit of life unicellular orgs
have only 1 cell multi have lots.

2. Mitosis – Type of cell division that results in 2
daughter cells having the same number and
kind of cromosomes as the parents nucleus

3. DNA – typical of ordiany tissuegroth conparewith

[Mitosis]

4. Genes –

5. Chromosomes –

6. Classification –

7. Kingdom –

8. Species –

9. Nutrient – _____

_____

_____

10. Food Chain – _____

_____

_____

11. Food Web – _____

_____

_____

12. Ecosystem – _____

_____

_____

13. Ecology – _____

_____

_____

14. Habitat – _____

_____

_____

Student Assignment Sheet Week 1
Cells

Experiment: What do plant cells look like?
    Materials
- ✓ Microscope
- ✓ Slide
- ✓ Onion skin
- ✓ Celery stalk

    Procedure

Note—*If you have never used a microscope before, please ask the teacher to demonstrate how to use a microscope before beginning this experiment.*

1. Read the introduction to this experiment and write a description of what you think you will see.
2. Make a wet mount slide of the onion skin. Look at the slide under the microscope on low power (100x) and then on high power (400x). Draw what you see for each.
3. Next, make a wet mount slide of the celery stalk. Look at the slide under the microscope on low power (100x) and then on high power (400x). Draw what you see for each.
4. Complete the experiment sheet.

Vocabulary & Memory Work
- ⬚ Vocabulary: cell, mitosis
- ⬚ Memory Work—This week, work on memorizing the Five Kingdoms and their Basic Characteristics.
    1. **Monerans** – Microscopic organisms that have a simple, single cell. (*Example: Bacteria*)
    2. **Protists** – A variety of complex, but mainly single-celled organisms. (*Example: Algae*)
    3. **Fungi** – Organisms that absorb food and reproduce by making spores. (*Example: Molds*)
    4. **Plants** – Living things that have many cells and most carry out photosynthesis. (*Example: Trees*)
    5. **Animals** – Organisms made up of many cells and live by eating food. (*Example: Humans*)

Sketch: Plant and Animal Cells
- ▦ Label the following on the Plant cell – cell wall, cell membrane, cytoplasm, vacuole, nucleus, chloroplasts
- ▦ Label the following on the Animal Cell – nucleus, nucleolus, mitochondria, vacuole, endoplasmic reticulum, ribosome, cell membrane, lysosome, cytoplasm

Writing
- ᗷ Reading Assignment: *Usborne Science Encyclopedia* pp. 250-251 (Plant Cells), pp. 298-299 (Animals Cells)
- ᗷ Additional Research Readings
    - 📖 The parts of a cell and what they do (including organelles): *UIDS* pp. 238-240
    - 📖 Cells: *DKEN* pp. 20-21

Dates to Enter
- 🕐 1595 – Zacharias Jansenn built the first microscope.
- 🕐 1665 – The cell was first discovered and named by Robert Hooke.
- 🕐 1839 – Cell theory was developed by Matthias Schleiden and Theodor Schwann.

Sketch Assignment Week 1

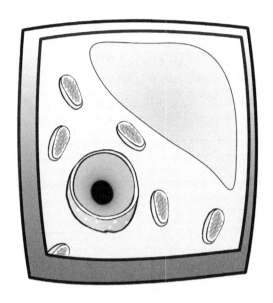

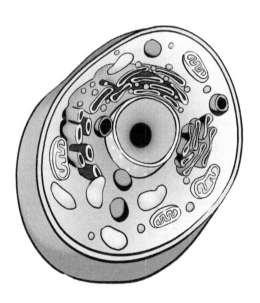

Experiment: What do plant cells look like?

Introduction

All living things are made up of tiny building blocks called cells. Cells carry out the necessary functions of life for the animal and plant. The best way to see these cells is by using a microscope. In this experiment, you will use a microscope to view the basic building blocks of celery and onion.

Hypothesis

I think that the cells will look like ___circles with stuff inside___

Materials

_____   _____

_____   _____

_____   _____

_____   _____

_____   _____

How to make a wet mount slide

1. Collect a thin slice of the sample and place it on the slide. (*Make sure the sample is very thin or else the cover slip will wobble and you won't get a very good view of the sample.*)
2. Place one drop of water over the sample. (*Make sure not to use too much water or else the cover slip will float away and again you won't be able to see the sample.*)
3. Place the cover slip at a 45 degree angle, with one edge touching the water, and let go. The slide is ready to be viewed.

Procedure

Observations

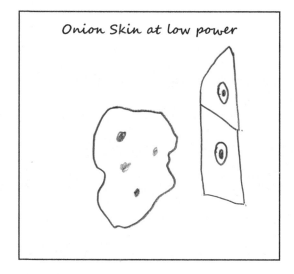

Celery Stalk at low power	Celery Stalk at high power

Conclusion

Written Assignment Week 1

Discussion Questions

Plant Cells, pp. 250-251

1. What are three of the basic parts of a plant cell and what does each do?
2. Are all plant cells the same? Why or why not?
3. What are the two stages of cell division in a plant cell? What happens in each of them?
4. What are the three types of plant tissue?

Animal Cells, pp. 298-299

1. What are three of the basic parts of an animal cell and what does each do?
2. Explain how cells come together to form organs.

_____

_____

_____

_____

_____

_____

_____

_____

_____

_____

_____

_____

_____

_____

_____

_____

_____

_____

_____

_____

Written Assignment Week 1

Student Assignment Sheet Week 2
DNA and Genetics

Experiment: DNA Extraction

Materials

- ✓ Banana Slices
- ✓ Dish soap
- ✓ Salt
- ✓ Ice-cold isopropyl alcohol (70% or higher)
- ✓ Zipper-style plastic bag
- ✓ Coffee filter
- ✓ Funnel
- ✓ Wooden coffee stirrer
- ✓ Test tube (or clear glass)

Procedure

1. Read the introduction to this experiment.
2. Make the extraction solution by measuring out ½ cup of water (120 mL). Add 2 tablespoons (35 mL) of liquid dish soap and 1 teaspoon (5 grams) of salt. Mix well.
3. Add several slices of banana to the baggie – about a third of a banana will do. Using your hands, squish the outside of the baggie to mash up the banana. Then, add 1 to 2 tablespoons (15 to 30 mL) of extraction solution. Keep squishing for about a minute or so to make sure that some of the DNA is released into the solution. (**Note**—*If you have time, you can let the mixture sit for ten to fifteen minutes so that even more DNA is released.*)
4. Line the funnel with the coffee filter and place the funnel in the test tube. Then, pour the contents of the baggie into the lined funnel and gently squeeze the liquid into the test tube. You want the tube to be about a quarter to a third full.
5. Slowly pour 2 to 3 tablespoons (30 to 45 mL) of the ice-cold alcohol into the test tube, so that you create a layer of alcohol resting on your fruit layer.
6. Watch what happens. Use the wooden coffee stirrer to gently stir the solution and collect any of the material you can see.
7. Draw conclusions and complete the experiment sheet.

Vocabulary & Memory Work

☐ Vocabulary: DNA, genes, chromosomes
☐ Memory Work—Continue to work on memorizing the Five Kingdoms and their Basic Characteristics.

Sketch: DNA to Chromosomes

▨ Label the following – DNA strand, Coiled chromatin, Chromosomes, Nucleus, Cell

Writing

෴ Reading Assignment: *Kingfisher Science Encyclopedia* pg. 135 Genes and Chromosomes
෴ Additional Research Readings
    📖 Genetics: *USE* pp. 380-381, *UIDS* pp. 324-325
    📖 Gene Therapy: *USE* pp. 382-383

Dates to Enter

🕑 1953 – James Watson and Francis Crick explain the double helix structure of DNA.

Sketch Assignment Week 2

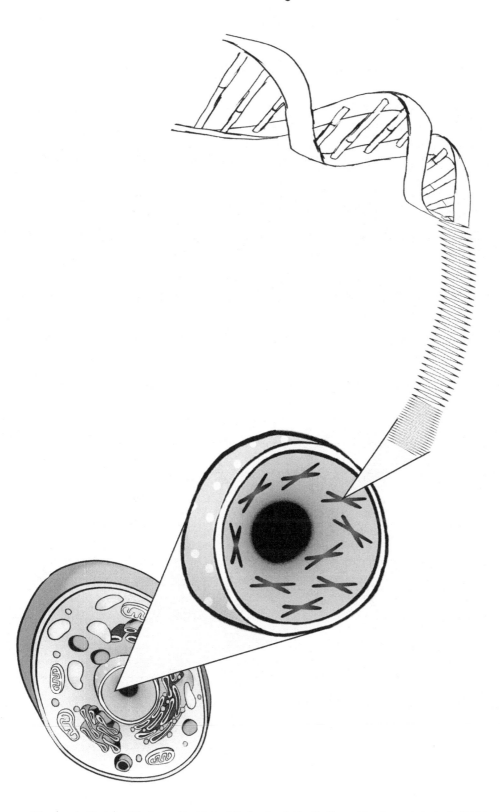

Experiment: DNA Extraction

Introduction

DNA is the stuff that tells our cells what to do and how to look. It resides in the nucleus of a cell, so as you can imagine, it is quite tiny. In fact, you normally need a very powerful microscope to see DNA for yourself. In today's experiment, you are going to try to extract DNA and cause it to join up into a mass you can see.

Materials

_____   _____

_____   _____

_____   _____

_____   _____

_____   _____

Procedure

_____

_____

_____

_____

_____

_____

_____

_____

_____

_____

_____

_____

_____

_____

_____

_____

Observations

Picture of what happened in the test tube

Results

_____

_____

_____

_____

_____

_____

_____

_____

_____

_____

_____

_____

Conclusion

_____

_____

_____

_____

_____

_____

_____

_____

_____

_____

_____

_____

_____

Written Assignment Week 2

Discussion Questions

1. What is the structure of DNA?
2. Where are chromosomes found and what do they contain?
3. What is DNA made up of?
4. How do genes always occur?
5. How are genes passed?

Written Assignment Week 2

Student Assignment Sheet Week 3
Classification

Experiment: What kind of tree is it?

Materials
- ✓ Dichotomous key for plants from this website:
    - 🌐 http://oregonstate.edu/trees/dichotomous_key/index.html
- ✓ Leaf from outside

Procedure
1. Go outside and choose a leaf from a tree that you want to identify.
2. Once inside, read the introduction to this experiment and then write down what kind of tree you think the leaf is from.
3. Use the dichotomous key from the website above to identify the leaf. (**Note**—*Although the website is for trees of the Pacific Northwest, it will help you identify trees from all over the US.*)
4. Complete the experiment sheet.

Vocabulary & Memory Work
- ☐ Vocabulary: classification, kingdom, species
- ☐ Memory Work—This week, continue working on the Five Kingdoms and their Basic Characteristics and add the Divisions of Life:

    Kingdom, Phylum, Class, Order, Family, Genus, Species
    (The following mnemonic can help you as you work on memorizing these:
    ***K**ing **P**hillip **C**an **O**nly **F**ind his **G**reen **S**hoes)*

Sketch: Divisions of Life
- 🖼 Label each box with Kingdom, Phylum, Class, Order, Family, Genus, Species

Writing
- 〰 Reading Assignment: *Usborne Science Encyclopedia* pp. 294-295 Classifying Plants, pp. 340-343 Classifying Animals
- 〰 Additional Research Reading
    - 📖 Classification: *DKEN* pp. 286-289
    - 📖 Classification of Living Things: *KSE* pp. 52-53
    - 📖 How Living Things are Classified: *DKEN* pp. 108-109

Dates to Enter
- 🕐 1707-1778 – Carolus (Carl) Linnaeus lived; he is credited with the current classification system.
- 🕐 1984 – The DNA profiling (fingerprinting) technique was first reported by Sir Alec Jeffreys at the University of Leicester in England.

Sketch Assignment Week 3

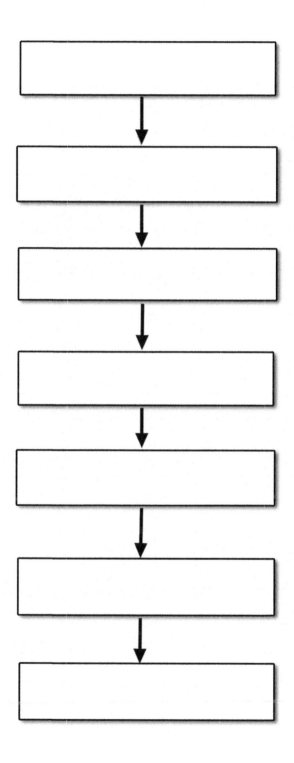

Experiment: What kind of tree is it?

Introduction

A dichotomous key is a method that can be used to identify a living thing, such as a tree. At each step, the user is given a question with two answers. The answers eventually lead to the identity of the living thing. In this experiment, you will be using a dichotomous key from the internet to identify an unknown leaf.

Hypothesis

I think the leaf is from the _____ tree.

Materials

_____   _____

_____   _____

_____   _____

_____   _____

_____   _____

Procedure

_____

_____

_____

_____

_____

_____

_____

_____

_____

_____

Results

_____

_____

_____

Observations

Drawing or Picture of my Leaf

Conclusion

_____

_____

_____

_____

_____

_____

_____

_____

_____

_____

_____

_____

Written Assignment Week 3

Discussion Questions

Classifying Plants, pp. 294-295

1. What are the five main kingdoms?
2. How do scientists classify living things?
3. What are the two main divisions of the plant kingdom?

Classifying Animals, pp. 340-343

1. What are the taxonomic ranks (or divisions of life)?
2. What language is a biological name given in and how is it created?

Written Assignment Week 3

Student Assignment Sheet Week 4
Nutrient Cycles

Experiment
☞ No experiment this week.

Sketch: Four Nutrient Cycles
- **Nitrogen Cycle** – Nitrogen gas in the atmosphere. Lightning combines nitrogen and oxygen which falls as weak nitric acid rain. Bacteria covert nitrogen compounds in the soil into nitrates. Bacteria take in nitrates in the soil and release nitrogen into the air. Plants take up nitrates in the soil. Animals get nitrogen from plants. Dead plants and animals release nitrogen. Bacteria convert the nitrogen compounds into nitrates.
- **Carbon Cycle** – Carbon dioxide in the atmosphere. Plants get carbon dioxide from the air. Animals get carbon from plants. Animals give off carbon dioxide when breathing, plants and animals die and their bodies decay. Fungi and bacteria give off carbon dioxide as they breakdown dead matter.
- **Phosphorus Cycle** – Phosphate-rich sediment turns into rock. Phosphate-rich rock erodes into small particles. Animals take in phosphates from plants. Phosphates are released into the soil from the remains of plants and animals, phosphates from fertilized fields are carried into the sea. Phosphates in marine organisms are broken down into marine sediment. Phosphate-rich sediment builds up on the sea bed.
- **Water Cycle** – The Sun's heat evaporates water from the Earth's oceans and lakes. Water vapor condenses to form clouds. Water vapor cools and falls as rain. Rainwater flows back into rivers, lakes, and seas.

Vocabulary & Memory Work
☐ Vocabulary: nutrient, food chain, food web, ecosystem
☐ Memory Work—Continue to work on memorizing the Divisions of Life and Five Kingdoms and their Basic Characteristics.

Writing
〰 Reading Assignment: *Usborne Science Encyclopedia* pp. 332-333 Food & Energy, pp. 334-335 Nutrient Cycle
〰 Additional Research Reading
　📖 Ecosystems: *UDIS* pp. 234-235
　📖 Food Chains and Webs: *DKEN* pp. 66-67
　📖 Nutrient Cycles: *DKEN* pp. 64-65

Dates to Enter
🕐 781-868 – Al-Jahiz lived; he is credited with the earliest description of a food chain.
🕐 1880 – The earliest graphic depiction of a food web is published by Lorenzo Camerano.

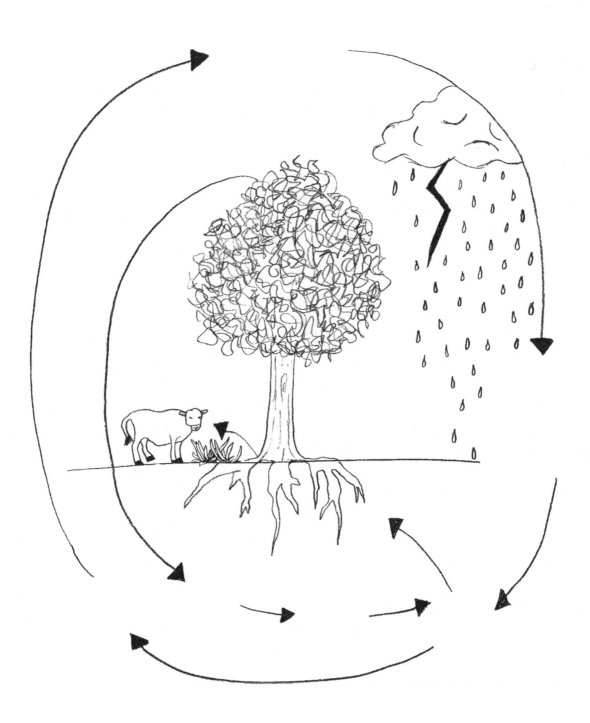

Sketch Assignment Week 4

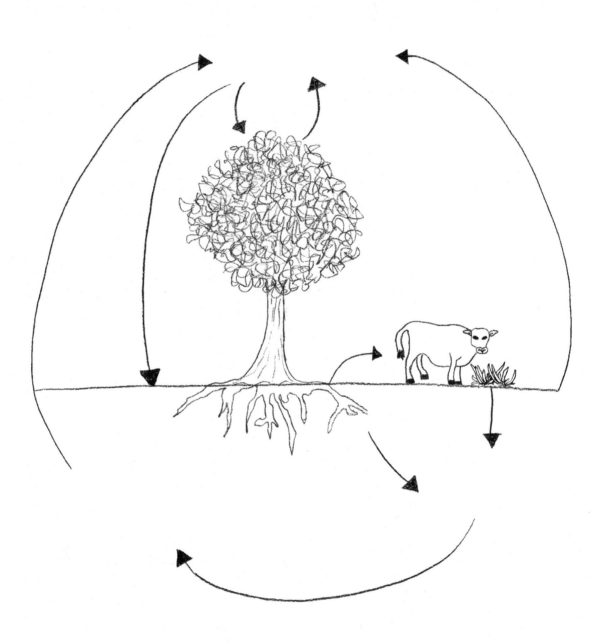

Sketch Assignment Week 4

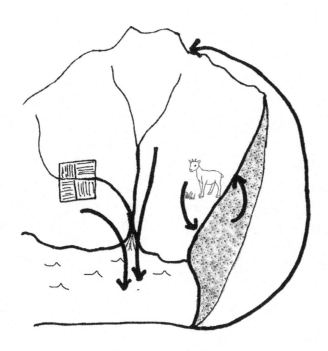

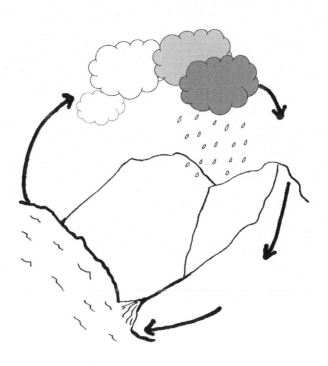

Written Assignment Week 4

Discussion Questions

Food and Energy, pp. 332-333

1. What is a food chain?
2. What are plants called in a food chain? What are animals called in a food chain?
3. What is the role of a decomposer? Name 3 types of decomposers.
4. What is a food web?

Nutrient Cycles, pp. 334-335

1. What does a living cycle (or nutrient cycle) do?
2. What are the three main living cycles (or nutrient cycles)?
3. What is pollution?

_____

_____

_____

_____

_____

_____

_____

_____

_____

_____

_____

_____

_____

_____

_____

_____

_____

_____

_____

_____

_____

Written Assignment Week 4

Student Assignment Sheet Week 5
Basic Ecology

Experiment: Habitat Diorama

    Materials
- ✓ Air dry clay
- ✓ Shoebox
- ✓ Paint
- ✓ Construction paper

    Procedure

1. Begin by choosing the habitat you would like to create (**Note**—*The grasslands, the rainforest, or the desert are a few options.*)
2. Make a brief sketch to plan out what you would like the habitat diorama to look like. Be sure to include some of the plants and animals that would be found in the habitat.
3. Get the shoebox and begin creating the background for the habitat. You can use paint or construction paper for this.
4. Using air dry clay, begin to make some of the plants and animals in the habitat. Then, place them in the habitat diorama. Put the finishing touches on the model and take a picture of it.
5. Fill out the Habitat Diorama Sheet.

Vocabulary & Memory Work

- ☐ Vocabulary: ecology, habitat
- ☐ Memory Work—Continue to work on memorizing the Divisions of Life and the Five Kingdoms and their basic characteristics.

Sketch: Ecological Succession

- 📷 Label the following – Pioneer community with mainly grasses and small insects, Shrubs and bushes begin to grow, Climax community with mature trees and a wide variety of animals

Writing

- ᕽ Reading Assignment: *Usborne Science Encyclopedia* pp. 330-331 Ecology
- ᕽ Additional Research Reading
  - 📖 Biomes and Habitats: *KSE* pp. 68-69
  - 📖 Living Things and their Environments: *UIDS* pp. 232-233

Dates to Enter

- 🕐 1866 – Ernst Haeckel first coined the term ecology.

Sketch Assignment Week 5

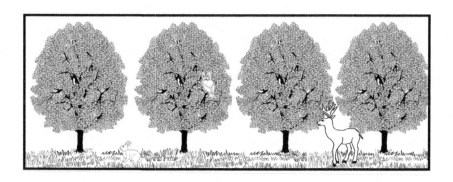

Habitat Diorama Sheet Week 5

Habitat Name: _____

Picture of my Habitat

What I learned from this project

_____
_____
_____
_____
_____
_____
_____

Written Assignment Week 5

Discussion Questions

1. What is a habitat?
2. What is a community?
3. What is an ecosystem?
4. What is an ecological niche?
5. What is a biome?
6. Name some of the world's biomes.

_____

_____

_____

_____

_____

_____

_____

_____

_____

_____

_____

_____

_____

_____

_____

_____

_____

_____

Written Assignment Week 5

Biology: Unit 2

Plants

Biology Unit 2: Plants
Vocabulary Sheet

Define the following terms as they are assigned on the Student Assignment Sheet.

1. Fungi – _____

   _____

   _____

2. Spore – _____

   _____

   _____

3. Yeast – _____

   _____

   _____

4. Algae – _____

   _____

   _____

5. Hydrophyte – _____

   _____

   _____

6. Photosynthesis – _____

   _____

   _____

7. Chlorophyll – _____

   _____

   _____

8. Frond – _____

   _____

   _____

9. Angiosperm – _____

10. Pollination – _____

_____

_____

11. Cotyledon – _____

_____

_____

12. Dicot – _____

_____

_____

13. Germination – _____

_____

_____

14. Monocot – _____

_____

_____

15. Gymnosperm – _____

_____

_____

16. Deciduous – _____

_____

_____

17. Evergreen – _____

_____

_____

Student Assignment Sheet Week 6
Fungi

Experiment: Can I grow mold?
> Materials
>> ✓ Bread
>> ✓ Plastic bag
>> ✓ Water
>
> Procedure
>> 1. Read the introduction to this experiment and then answer the question.
>> 2. Moisten the bread by sprinkling it with water.
>> 3. Place it in the plastic bag and set the bag in a warm, dark place.
>> 4. Observe the bread every day for five days. Each day, record your observations by taking a picture or by drawing what you see on the experiment sheet.
>> 5. After five days, make your final observations of the bread and finish the experiment sheet.
>> 6. Once you are done, throw the bread away. **DO NOT** open the plastic bag.

Vocabulary & Memory Work
> ☐ Vocabulary: fungi, spores, yeast
> ☐ Memory Work—There is no memory work for this week.

Sketch: Anatomy of a Fungus
> ▨ Label the following – fruiting body, gills, stalk, mycelium

Writing
> ᕫ Reading Assignment: *Usborne Science Encyclopedia* pp. 284-285 Fungi
> ᕫ Additional Research Reading
>> ▢ Fungi: *DKEN* pp. 114-115
>> ▢ Fungi & Lichens: *KSE* pg. 55

Dates to Enter
> 🕐 1588 – Giambattista della Porta first observes fungal spores.
> 🕐 1836 – English naturalists Miles Joseph Berkeley first used the word mycology to mean the study of fungi.
> 🕐 1928 – Alexander Fleming discovers penicillin.

Sketch Assignment Week 6

Experiment: Can I grow mold?

Introduction

A mold is neither a plant nor an animal. It is part of the Fungi Kingdom and is usually found in dark, damp places. A mold does not contain chlorophyll, so it cannot produce its own food from light like most other plants can. Instead, mold feeds on living or once-living matter. In this experiment, you will see if you can grow mold using a piece of bread.

Hypothesis

Can I grow mold?   Yes   No

Materials

_____   _____

_____   _____

_____   _____

_____   _____

_____   _____

Procedure

Observations

Bread after Day 1	Bread after Day 2

Bread after Day 3	Bread after Day 4

Conclusion

_____

_____

_____

_____

Written Assignment Week 6

Discussion Questions

1. What are fungi?
2. Explain the structure of a fungus.
3. What are molds and mildews and where are they found?
4. Where do fungi get their food?
5. Are fungi harmful or helpful? Explain why or why not.

_____

_____

_____

_____

_____

_____

_____

_____

_____

_____

_____

_____

_____

_____

_____

_____

_____

_____

_____

_____

_____

Written Assignment Week 6

Student Assignment Sheet Week 7
Simple Plants

Experiment: Can I grow algae?

Materials
- ✓ Pond or aquarium water
- ✓ Small glass jar
- ✓ Eye dropper
- ✓ Microscope
- ✓ 2 Slides and cover slips

Procedure
1. Read the introduction to this experiment and then answer the question.
2. Collect some pond or aquarium water in a small glass jar.
3. Place a drop of water onto a slide and cover it with a cover slip, using the same method you learned in week one.
4. Look at it under the microscope using the 10x objective lens. Move the slide around slowly to see if you can find any green algal cells.
5. Set the jar out on a sunny window sill and observe what happens in the jar over five days.
6. After five days, look at the water under the microscope again using the same procedure you used in steps three and four. Then, complete the experiment sheet.

Vocabulary & Memory Work
- ☐ Vocabulary: algae, hydrophyte, photosynthesis
- ☐ Memory Work—This week, work on memorizing the photosynthesis equation.
  Carbon Dioxide + Water + Energy from the Sun ⟶ Carbohydrates + Oxygen

Sketch: Divisions of Plants
- ▥ Read *Usborne Science Encyclopedia* pg. 295 The Plant Kingdom.
- ▥ Label the following: nonvascular plants, vascular plants, plants without seeds, plants with seeds, gymnosperms, angiosperms, monocots, dicots

Writing
- ᘒ Reading Assignment: *Usborne Science Encyclopedia* pg. 264 Plant Food (Photosynthesis section), pp. 281-282 Water Plants
- ᘒ Additional Research Reading
  - 📖 Photosynthesis: *UIDS* pp. 254-255, *DKEN* pp. 24-25
  - 📖 Plant Anatomy: *KSE* pp. 56-57

Dates to Enter
- ☼ 2600 BC- 2000 BC – Ancient Egyptians used different colors of algae as eye makeup.
- ☼ 1811-1866 – William Harvey was the first to divide algae into four divisions based on their pigmentation.

Sketch Assignment Week 7

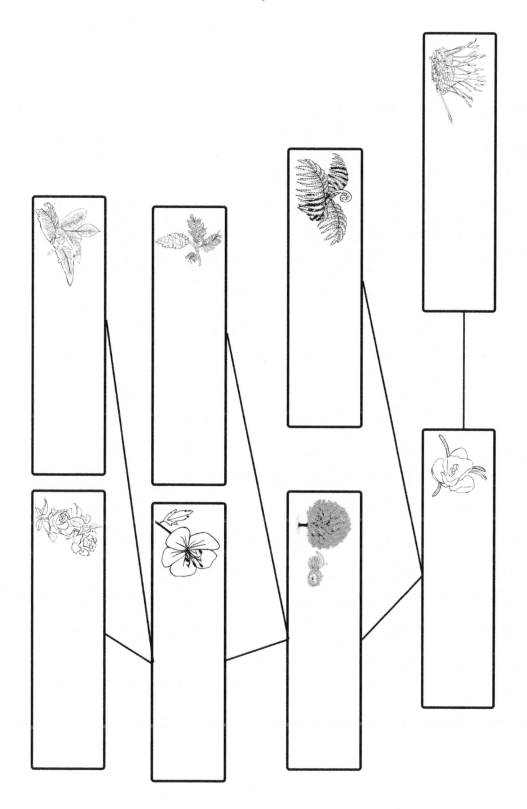

Experiment: Can I grow algae?

Introduction

Green algae are the most diverse category of algae. There are over seven thousand different species that grow in many types of aquatic habitats. Green algae have chlorophyll, which captures light energy and uses it to produce sugars for food. In this experiment, you will be attempting to grow green algae from water from a pond or an aquarium.

Hypothesis

Can I grow algae?   Yes   No

Materials

_____   _____

_____   _____

_____   _____

_____   _____

_____   _____

Procedure

_____

_____

_____

_____

_____

_____

_____

_____

_____

_____

Observations and Results

Water on Day 1	Water on Day 5

_____

_____

_____

_____

_____

_____

Conclusion

_____

_____

_____

_____

_____

_____

_____

_____

_____

Written Assignment Week 7

Discussion Questions

Plant Food (Photosynthesis section), pg. 264

1. What is an autotrophic plant?
2. Where does photosynthesis take place?
3. What is chlorophyll, where is it, and what does it do?

Water Plants, pp. 281-282

1. What are the two categories of water plants? Explain a little about each category.
2. Where are algae typically found?
3. How are algae different from other plants? How are they similar?
4. What is seaweed?
5. What is eutrophication?

Written Assignment Week 7

Student Assignment Sheet Week 8
Flowerless Plants

Experiment: Observation of a Fern
  Materials
  - ✓ Fern frond (with spores if possible)
  - ✓ Magnifying glass
  - ✓ Microscope
  - ✓ Slide & cover slip

  Procedure
  1. Read the introduction to this experiment.
  2. Observe the fern using your eyes and the magnifying glass, and then answer the questions on the experiment sheet.
  3. Make a wet mount slide using the directions from the experiment in week one. Look at the slide using the 40x objective, then draw what you see on the experiment sheet.
  4. Draw conclusions and complete the experiment sheet.

Vocabulary & Memory Work
  - ☐ Vocabulary: chlorophyll, frond
  - ☐ Memory Work—Continue to work on the photosynthesis equation.

Sketch: Life Cycle of a Fern
  - ▣ Label the following – The mature plant produces sori, which contain clusters of sporangia. Sporangia release spores into the air. Spores grow into the prothallus, which produces the sex cells. The male and female sex cells fuse. The resulting cell grows into a fern plant.

Writing
  - ᨓ Reading Assignment: *Usborne Science Encyclopedia* pp. 282-283 Flowerless Plants
  - ᨓ Additional Research Reading
    - 📖 Non-flowering Plants: *KSE* pg. 58
    - 📖 Flowerless Plants: *DKEN* pp. 116-117
    - 📖 Leaves: *UIDS* pp. 248-250

Dates to Enter
  - 🕐 1855 – Charles Kingsley coins the term Pteridomania or Fern Craze to describe the Victorian era craze of fern collecting and decorating with the fern motif in pottery and glass.

Sketch Assignment Week 8

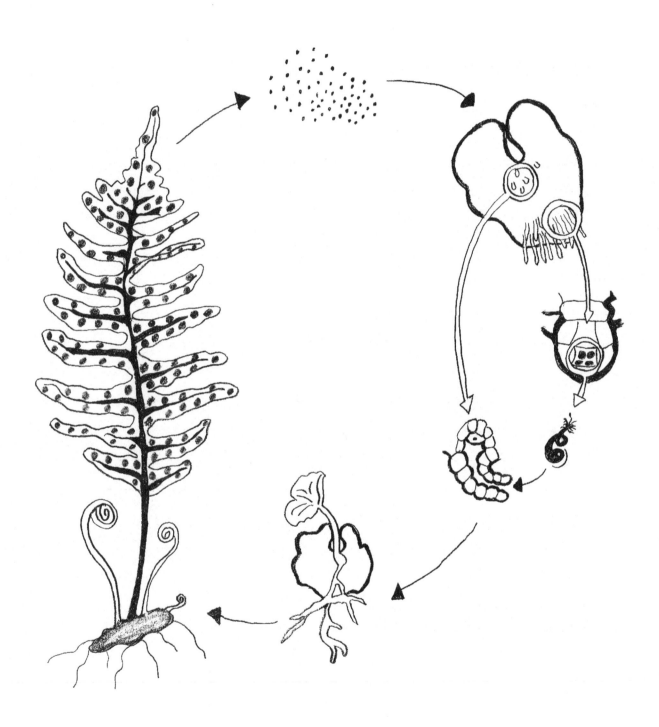

Experiment: Observation of a Fern

Introduction

Ferns are primitive plants that are generally found in humid forests and near river banks. Ferns do not flower. They also release spores instead of seeds. In this experiment, you will examine and observe the structure of a fern.

Materials

_____   _____

_____   _____

_____   _____

_____   _____

_____   _____

Procedure

_____

_____

_____

_____

_____

_____

_____

_____

_____

_____

_____

Observations and Results

How does the fern smell and feel?

_____

_____

_____

_____

Describe the basic structure of the fern.

_____

_____

_____

_____

_____

How the fern looked under the magnifying glass	How the fern looked under the microscope

Conclusion

_____

_____

_____

_____

_____

_____

_____

_____

Written Assignment Week 8

Discussion Questions

1. What is asexual reproduction?
2. How do liverworts and mosses get the water they need to survive?
3. Where are ferns found?
4. Describe how a fern grows.
5. What does "alternation of generations" mean?
6. How do ferns reproduce?

Written Assignment Week 8

Student Assignment Page Week 9
Flowering Plants

Experiment: Dissection of a Flower

> Materials
> - ✓ Flower (either lily, poppy or tulip)
> - ✓ Magnifying glass
> - ✓ Microscope
> - ✓ 2 Slides and a cover slip
> - ✓ Razor blade

> **⚠ CAUTION**
>
> Be sure to have the teacher assist you with the razor blade used for steps through five of this experiment!

> Procedure
> 1. Read the introduction to this experiment.
> 2. Observe the flower using your eyes and the magnifying glass, and then answer the questions on the experiment sheet.
> 3. Using a razor blade, remove the stamen. (**Note** – *Use the diagram on the experiment sheet to assist you in identifying the various parts.*)
> 4. Gently wipe some of the pollen onto a slide, examine it under a microscope, and draw what you see. Then cut open the anther and use a magnifying glass to take a closer look.
> 5. Next, remove the pistol. Cut it in half and use a magnifying glass to take a closer look. Try to locate the ovary, stigma, and style of the flower.
> 6. Finally, use a razor blade to cut a thin cross section of the stem. Then, using the directions from week one, make a wet mount slide. Look at the slide under the microscope, draw what you see, and complete the experiment sheet.

Vocabulary & Memory Work

- ☐ Vocabulary: angiosperm, pollination
- ☐ Memory Work—Begin to memorize the Parts of a Flowering Plant.
  1. **Root** – It helps to anchor the plant and absorb nutrients.
  2. **Stem** – It holds the plant up and serves as the transport system for the plant.
  3. **Leaf** – It absorbs sunlight and produces energy for the plant through chlorophyll.
  4. **Flower** – It is the reproductive part of the plant.
  5. **Seed** – It contains the material necessary to grow a new plant.

Sketch: Parts of a Flower

- 🖼 Label the following – Petals, Sepals, Pistil, Stigma, Style, Ovary, Ovules, Stamen, Anther, and Filament

Writing

- ᕕ Reading Assignment: *Usborne Science Encyclopedia* pp. 270-273 Flowering Plants
- ᕕ Additional Research Reading
  - 📖 Flowering Plants: *KSE* pp. 59-61, *DKEN* pp. 126-137
  - 📖 Flowers: *UIDS* pp. 256-257

Dates to Enter

- 🕐 77 – Roman naturalist, Pliny the Elder, completes *Historia Naturalis*, which is the first encyclopedia of nature.
- 🕐 1872 – Yellowstone National Park is created in the United States. It is the first park created for the preservation of a natural environment.

Sketch Assignment Week 9

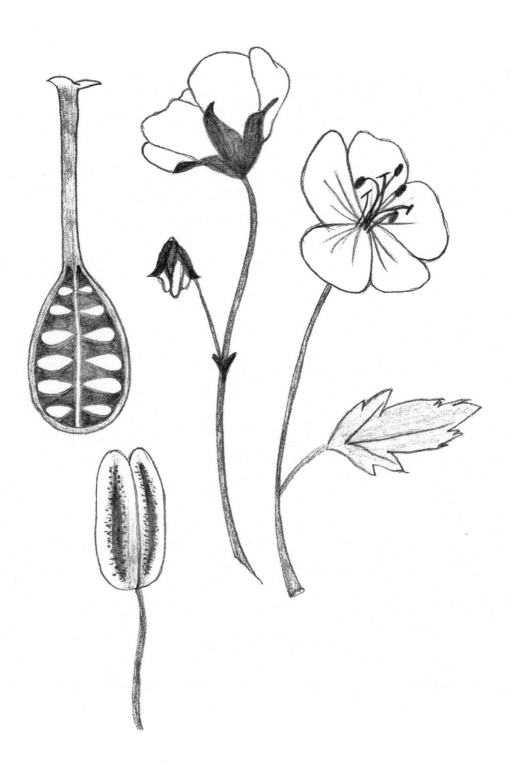

Experiment: Dissection of a Flower

Introduction

Flowers are the reproductive part of the plant. They contain four main sets of organs; which are the sepals, the petals, the stamens, and the pistil. The sepals and petals are there to attract insects for pollination. The stamen and pistil are the reproductive organs of the flower. In this experiment, you will examine a single flower close up.

Materials

_____   _____

_____   _____

_____   _____

_____   _____

_____   _____

Procedure

_____

_____

_____

_____

_____

_____

_____

_____

_____

Observations and Results
How does the flower smell and feel?

_____

_____

_____

_____

Describe the general appearance of the flower.

_____

_____

_____

_____

_____

_____

_____

_____

Pollen under a micrscope	*Cross-section of the stem under a micrscope*

Conclusion

_____

_____

_____

_____

_____

_____

Written Assignment Week 9

Discussion Questions

1. What are several characteristics common to all flowering plants?
2. Explain the purpose that each of the following parts of a flowering plant serve – bud, sepals, petals, necatries.
3. Whare are the male parts of a flower?
4. What are the female parts of a flower?
5. Briefly describe plant fertilization.
6. Explain two ways a flower can be pollinated.
7. What is the difference between animal and wind pollination?

Written Assignment Week 9

Student Assignment Sheet Week 10
Seeds and Fruit

Experiment: How many leaves will sprout?

Materials
- ✓ Corn seed
- ✓ Bean seed
- ✓ Soil
- ✓ 2 paper cups

Procedure
1. Read the introduction to this experiment and answer the question.
2. Fill each of the cups two-thirds of the way with soil. Label one cup "corn seed" and the other "bean seed."
3. Gently press each of the seeds into the dirt in the labeled cup. Water each seed each time the soil feels dry as you wait for it to sprout.
4. Once the seeds sprout, write down how many leaves sprouted and finish filling out the experiment sheet.

Vocabulary & Memory Work
- ☐ Vocabulary: cotyledon, dicot, germination, monocot
- ☐ Memory Work—Begin to work on the Parts of a Flower and be able to point out where they are. Use the list from the Sketch Assignment for the Memory Work.

Sketch: Parts of a Bean Seed
- ▣ Label the following – testa, hilum, micropyle, outside of seed, plumule, radicle, food store, cross-section of seed

Writing
- ᘓ Reading Assignment: *Usborne Science Encyclopedia* pp. 274-277 Seeds and Fruit
- ᘓ Additional Research Reading
  - 📖 Fruits and Seeds: *KSE* pp. 62-63,
  - 📖 Flowers and Seeds: *DKEN* pp. 128-129
  - 📖 Seeds: *UIDS* pp. 260-261

Dates to Enter
- 🕐 1865 – Austrian monk Gregor Mendel demonstrates the principles of heredity using pea plants.

Sketch Assignment Week 10

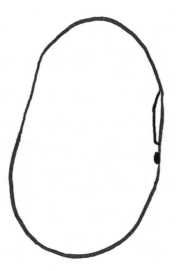

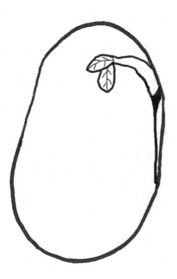

Experiment: How many leaves will sprout?

Introduction

Every seed contains a plant embryo, a food store, and an outer coat for protection. When the right temperature, amount of light and amount of moisture are reached, the seed will germinate. The first leaves to grow are called the cotyledons or seed leaves. In this experiment, you will see whether your seeds are monocots (has one seed leaf) or dicots (have two seed leaves).

Hypothesis

I think that my corn seed will sprout _____ leaves.

I think that my bean seed will sprout _____ leaves.

Materials

_____	_____
_____	_____
_____	_____
_____	_____
_____	_____

Procedure

_____

_____

_____

_____

_____

_____

_____

_____

_____

_____

Observations and Results

Picture of my corn seed	*Picture of my bean seed*

My corn seed sprouted _____ leaves.   My bean seed sprouted _____ leaves

_____

_____

_____

_____

_____

Conclusion

_____

_____

_____

_____

_____

_____

_____

_____

Written Assignment Week 10

Discussion Questions

1. What is the purpose of fruit? Where does it come from?
2. What are succulent fruits? Name the three key types of succelents fruits and describe each.
3. How do dry fruits differ from succulent fruits?
4. Name 3 ways that seeds can be dispersed.
5. What does a seed need to germinate?
6. What is a monocot? Name some monocotyledons (monocots).
7. What is a dicot? Name some dicotyledons (dicots).

_____

_____

_____

_____

_____

_____

_____

_____

_____

_____

_____

_____

_____

_____

_____

_____

_____

_____

_____

_____

_____

_____

Written Assignment Week 10

Student Assignment Sheet Week 11
Broadleaf Trees

Experiment: How tall and how old is this tree?
  Materials
  - ✓ Ruler
  - ✓ String
  - ✓ Measuring tape
  - ✓ Partner
  - ✓ Broadleaf Tree

  Procedure
  1. Read the introduction to this experiment and answer the questions.
  2. Estimate the height of the tree:
       - ↻ Stand near the tree, holding a ruler vertically in front of your face.
       - ↻ Carefully walk backwards until the ruler appears to be the same height as the tree.
       - ↻ Have your partner measure the distance from you to the tree and record this on the experiment sheet. This is the approximate height of the tree. (**Note**—*Be sure to include the units.*)
  3. Estimate the age of the tree:
       - ↻ Measure from the ground five feet up the trunk of the tree.
       - ↻ Next, use a string to measure around the tree.
       - ↻ Then, take the string and measure how many inches long it is. If the tree measures twenty inches around, it is around twenty years old. (**Note**—*This method is not always accurate for all types of trees, but it will help you to estimate the age of the tree.*)
  4. Complete the experiment sheet.

Vocabulary & Memory Work
  - ☐ Vocabulary: gymnosperm, deciduous, evergreen
  - ☐ Memory Work—Continue to work on the Parts of a Plant and the Parts of a Flower.

Sketch: Anatomy of a Broadleaf Tree
  - ▨ Label the following – crown, trunk, leaves, flower, fruit, roots

Writing
  - ๛ Reading Assignment: *Kingfisher Science Encyclopedia* pp. 64-65 Trees
  - ๛ Additional Research Reading
       - ▥ Older Plants: *USE* pp. 256-257 or *UIDS* pp. 246-257
       - ▥ Conifers: *DKEN* pp. 130-131
       - ▥ Broad-leaved Trees: *DKEN* pp. 132-133

Dates to Enter:
  - ☉ 1968 – Redwood National Park is created, home of the world's tallest tree which is 115.5 m (379.1 ft) tall.

Student Guide Biology Unit 2 Plants ~ Week 11 Trees

Sketch Assignment Week 11

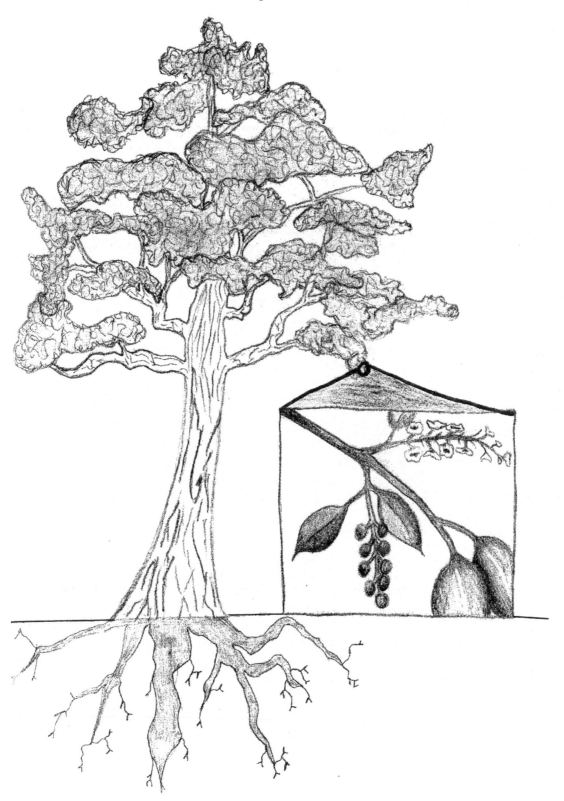

Experiment: How tall and how old is my tree?

Introduction

Broadleaf trees are the most abundant trees in the world. There are thousands of varieties, but all have broad, flat leaves. Broadleaf trees belong to a group of flowering plants called angiosperms. These types of trees dominate many of the world's forests. In this experiment, you are going to estimate the height and age of a broadleaf tree in your area.

Hypothesis

I think that my tree is _____ tall. (*Be sure to include units.*)

I think that my tree is _____ years old.

Materials

_____   _____

_____   _____

_____   _____

_____   _____

_____   _____

Procedure

_____

_____

_____

_____

_____

_____

_____

_____

_____

_____

_____

Observations and Results

I estimate my tree to be _____ tall. (*Be sure to include units.*)

I estimate my tree to be _____ years old.

Picture of the tree I measured

Conclusion

_____

_____

_____

_____

_____

_____

_____

_____

Written Assignment Week 11

Discussion Questions

1. What is the main characteristic of coniferous trees? What are the benefits of these characteristics?
2. What are the main characteristics of broadleaf trees?
3. What changes happen in a leaf in the fall?
4. What is one difference between broad-leaved and coniferous trees?
5. What feature of a tree allows it to grow taller than other plants?
6. Name two ways that trees are beneficial to the environment.

_____

_____

_____

_____

_____

_____

_____

_____

_____

_____

_____

_____

_____

_____

_____

_____

_____

_____

_____

_____

Written Assignment Week 11

Biology: Unit 3

Invertebrates

Biology Unit 3: Invertebrates
Vocabulary Sheet

Define the following terms as they are assigned on the Student Assignment Sheet.

1. Invertebrate – _____

_____

_____

2. Coral – _____

_____

_____

3. Echinoderm – _____

_____

_____

4. Medusa – _____

_____

_____

5. Polyp – _____

_____

_____

6. Mollusk – _____

_____

_____

7. Bivalve – _____

_____

_____

8. Crustacean – _____

_____

_____

9. Larva – _____

_____

_____

10. Exoskeleton – _____

_____

_____

11. Antennae – _____

_____

_____

12. Life Cycle – _____

_____

_____

13. Metamorphosis – _____

_____

_____

14. Nymph – _____

_____

_____

15. Thorax – _____

_____

_____

Student Assignment Sheet Week 12
Annelids, Platyhelminthes, Nematodes

Experiment: Can worms mix the soil?

Materials
- ✓ Small plastic soda bottle
- ✓ Sand, Loam (potting soil), Crushed leaves and/or plant matter
- ✓ Worms (2-3, can collect from outside)
- ✓ Water, Paper towel, Rubber band

Procedure
1. Read the introduction to this experiment and answer the question.
2. Layer ¼ cup (60 mL) each of sand, loam, and crushed leaves. Repeat one time.
3. Then, add ¼ cup (60 mL) of water if the sand, loam, and crushed leaves are dry.
4. Add the worms and cover the bottle with a paper towel. Use the rubber band to secure it in place. Then, let the bottle sit in a cool dark place for three days.
5. After three days, observe what has happened and complete the experiment sheet.

Vocabulary & Memory Work
- ☐ Vocabulary: invertebrate
- ☐ Memory Work—Begin to work on memorizing the Basic Phyla of the Animal Kingdom and their characteristics. This week work on:
  1. **Annelids** – Animals that are worm-like and have segmented bodies (*Example: Earthworms*).
  2. **Flatworms** (or Platyhelminthes) – Animals that are worm-like and have flat, unsegmented bodies (*Example: Flatworms*).
  3. **Nematodes** – Animals that have round worm-like bodies with no segments (*Example: Roundworms*).

Sketch: Anatomy of a Worm
- 🔳 Label it with the following – casts of undigested soil, body is made of ring-like segments, clitelum produces cases for worm eggs, head

Writing
- ∽ Reading Assignment: *Kingfisher Science Encyclopedia* pg. 72 Worms
- ∽ Additional Research Reading
  - 📖 Worms: *DKEN* pp, 144-145

Dates to Enter
- ⏲ 30-200 – Accounts of the use of leeches for the medically dubious practice of blood-letting came from China around 30 AD, India around 200 AD, ancient Rome around 50 AD, and later throughout Europe.
- ⏲ 1881 – Charles Darwin's book *The Formation of Vegetable Mould through the Action of Worms* presented the first scientific analysis of earthworms' contributions to soil fertility.

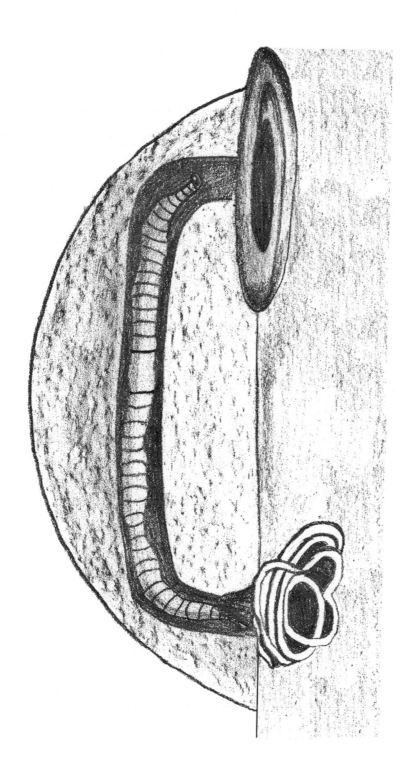

Experiment: Can worms mix soil?

Introduction

The earthworm is found in almost every habitat all over the world. It lives underground and feeds on broken down plant matter. As the earthworm moves through the soil, it takes in the organic matter, processes it in its gut and then ejects it out the other end as waste. In this experiment, you will determine whether or not earthworms can mix the soil they move through.

Hypothesis

Can worms mix soil?   Yes   No

Materials

_____   _____

_____   _____

_____   _____

_____   _____

_____

Procedure

_____

_____

_____

_____

_____

_____

_____

_____

_____

_____

_____

_____

Observations

Picture of my bottle before	Picture of my bottle after

_____

_____

_____

_____

Conclusion

_____

_____

_____

_____

_____

_____

Written Assignment Week 12

Discussion Questions

1. Name two places worms can be found.
2. What are the four main groups of worms?
3. Where do parasitic worms live?
4. Describe what earthworms do.
5. What is an earthworm's body made up of?
6. How do most marine worms eat?

_____

Written Assignment Week 12

Student Assignment Sheet Week 13
Cnidarians and Echinoderms

Experiment: How does coral grow?

Materials
- ✓ Plastic bowl
- ✓ Porous material (such as a sponge, brick or charcoal)
- ✓ Water
- ✓ Salt
- ✓ Liquid bluing
- ✓ Measuring spoons

Procedure
1. Read the introduction to this experiment and answer the question.
2. On day one, place the porous material in the bottom of the bowl. Mix together 2 TBSP (30 mL) each of water, salt, and liquid bluing and pour them over the porous material. Set the bowl where it won't be disturbed, but will still have good air flow.
3. On day two, add two more tablespoons of salt.
4. On day three, pour 2 TBSP (30 mL) each of water, salt, and liquid bluing into the bowl, but not directly over the porous material. Crystal formations should be beginning to appear, if not add 2 TBSP (30 mL) of ammonia to the bowl.
5. After four days, observe what has happened and complete the experiment sheet.

Vocabulary & Memory Work
- ☐ Vocabulary: coral, echinoderm, medusa, polyp
- ☐ Memory Work—Continue to work on memorizing the Basic Phyla of the Animal Kingdom and their characteristics. This week add:

  4. Cnidarians – Animals that live in water and have sack-like bodies with a single opening (*Example: Jellyfish*).

  5. Echinoderms – Animals with spiny skin, sucker feet, and a five-rayed body (*Example: Starfish*).

Sketch: Anatomy of a starfish
- Label the following – arm, aboral surface, each arm is covered with tube feet on the oral side

Writing
- ☞ Reading Assignment: *Kingfisher Science Encyclopedia* pg. 70 Marine Invertebrates
- ☞ Additional Research Reading
  - ☐ Jellyfish and Coral: *DKEN* pp. 146-147
  - ☐ Coral Reef: *DKEN* pp. 72-73

Dates to Enter
- ☐ No dates to be entered this week.

Sketch Assignment Week 13

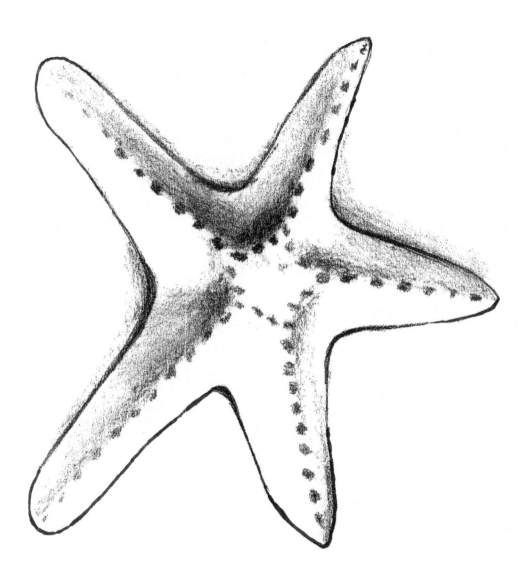

Experiment: How does coral grow?

Introduction

Coral reefs are found in many of the world's oceans. The most famous coral reef is the Great Barrier Reef, just off the coast of Australia. Coral Reefs are made up of thousands of tiny animals called coral. These sea animals have soft bodies and use their stinging tentacles to catch food as it swims by. In order to protect their soft bodies, corals create a stony cup around themselves that they can live in. The stony cups of many corals can join together to form a coral reef. In this experiment, you will determine the method by which corals form their stony cups.

Hypothesis

Coral grows by _____

_____

_____

_____

Materials

_____   _____

_____   _____

_____   _____

_____   _____

_____   _____

Procedure

_____

_____

_____

_____

_____

_____

_____

_____

_____

_____

_____

_____

Observations & Results

Picture of my crystals

_____

_____

_____

Conclusion

_____

_____

_____

_____

_____

_____

_____

_____

Written Assignment Week 13

Discussion Questions

1. What are invertebrates?
2. Are there more invertebrates or vertebrates in the ocean?
3. How do corals grow?
4. How do corals and algae work together?
5. What is the purpose of a jellyfish's tentacles?
6. What are two common characteristics of echinoderms? Name a few examples of echinoderms.

_____

_____

_____

_____

_____

_____

_____

_____

_____

_____

_____

_____

_____

_____

_____

_____

_____

_____

_____

Written Assignment Week 13

Student Assignment Sheet Week 14
Mollusks

Experiment: Do snails prefer caffeine?

Materials
- ✓ Snail
- ✓ 2 Lettuce leaves
- ✓ Caffeinated drink (preferably black coffee)
- ✓ Empty milk jug
- ✓ Paper towel
- ✓ Rubber band

Procedure
1. Read the introduction to this experiment and answer the question.
2. Cut off the top of the milk jug so that you can reach into it. Then dip one leaf into the caffeinated drink and leave one plain. Place each leaf on opposite sides of the milk jug.
3. Place the snail in the middle of the milk jug. Cover with the paper towel, secure with the rubber band, and set the milk jug aside.
4. After several days, observe which leaf was eaten and complete the experiment sheet.

Vocabulary & Memory Work
- ☐ Vocabulary: mollusk, bivalve
- ☐ Memory Work—Continue to work on memorizing the Basic Phyla of the Animal Kingdom and their characteristics. This week add:
  5. Mollusks – Animals with soft-bodies; most have shells (*Example: Snails*).
  6. Porifera – Animals that have perforated interior walls; most feed on bacteria (*Example: Sponges*).

Sketch: Anatomy of a snail
- ▨ Label the following – chalky shell, mantle, soft body

Writing
- ✎ Reading Assignment: *Kingfisher Science Encyclopedia* pg. 71 Mollusks
- ✎ Additional Research Reading
  - 📖 Snails and Slugs: *DKEN* pp. 148-149
  - 📖 Bivalves: *DKEN* pp. 150-151
  - 📖 Octopus and Squid: *DKEN* pp. 152-153

Dates to Enter
- 🕐 1795 – In the scientific literature, gastropods were described under the vernacular (French) name "gasteropodes" by Georges Cuvier.

Sketch Assignment Week 14

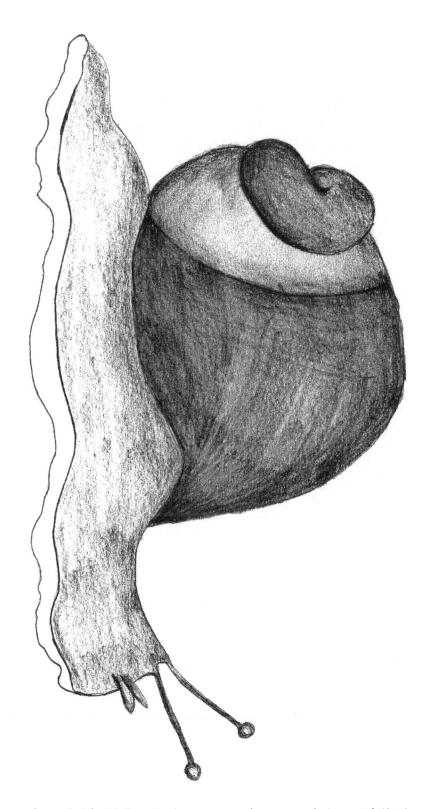

Experiment: Do snails prefer caffeine?

Introduction

Snails are a part of the mollusk phyla and the gastropod class. All gastropods have soft bodies that are protected by a hard shell. They also have a muscular foot to help them move along and a tongue that is covered in tiny teeth. Snails use their tongue to scrape up their food as they move by. In this experiment, you will determine whether snails prefer plain lettuce or lettuce that has been laced with caffeine.

Hypothesis

The snail will eat more of the _____ lettuce leaf.

Materials

_____   _____

_____   _____

_____   _____

_____   _____

_____   _____

Procedure

_____

_____

_____

_____

_____

_____

_____

_____

_____

_____

_____

Observations and Results

_____

_____

_____

_____

_____

_____

_____

_____

_____

_____

_____

_____

_____

Picture of my set-up

Conclusion

_____

_____

_____

_____

_____

_____

_____

_____

_____

Written Assignment Week 14

Discussion Questions

1. What are the characteristics of a mollusk?
2. What animals are included in the mollusk phyla and where are they found?
3. Describe a bivalve.
4. What distinguishes octopuses and squids from the rest of the mollusk phyla?
5. How do octopuses and squids move through the water?
6. What is the difference between slugs and snails?
7. What do snails use their iron teeth for?

Written Assignment Week 14

Student Assignment Sheet Week 15
Crustaceans (Arthropods)

Experiment: Can I dissolve shrimp shells?
Materials
- ✓ 3 Glass cups
- ✓ 3 Uncooked shrimp shells
- ✓ Bleach
- ✓ White vinegar
- ✓ Salt & water
- ✓ Gloves

Procedure
1. Read the introduction to this experiment and answer the questions.
2. Begin by labeling the cups #1, 2, and 3. Mix 2 TBSP (28 g) of salt with 1 cup (240 mL) of hot water; pour it into cup #1.
3. Mix 1 TBSP (15 mL) bleach with 1 cup (240 mL) of water; pour into cup #2.
4. Mix 1 TBSP (15 mL) vinegar with and 1 cup (240 mL) of water; pour into cup #3.
5. Place one shrimp shell in each cup; let the cups sit for three hours.
6. Put on the gloves and remove each of the shells. Thoroughly rinse & observe the changes.
7. Complete the experiment sheet.

Vocabulary & Memory Work
- ☐ Vocabulary: crustacean, larva, exoskeleton
- ☐ Memory Work—Continue to work on memorizing the Basic Phyla of the Animal Kingdom and their characteristics. This week add:
  - **8. Arthropods** – Animals that have segmented bodies, jointed legs; most have a hard exoskeleton (*Example: Crabs*).

Sketch: Anatomy of a crustacean
- 🖾 Label the following – antenna, pincer, compound eye, jointed leg, exoskeleton

Writing
- ᕫ Reading Assignment: *Kingfisher Science Encyclopedia* pg. 73 Crustaceans
- ᕫ Additional Research Reading
  - 📖 Crustaceans: *DKEN* pp. 158-159
  - 📖 Arthropods: *DKEN* pp. 156-157

Dates to Enter
- 🕐 1772 – The earliest valid work to use the name "Crustacea" was Morten Thrane Brünnich's *Zoologiæ Fundamenta*.

Sketch Assignment Week 15

Experiment: Can I dissolve shrimp shells?

Introduction

Shrimps are part of the arthropod phyla and the class crustacean. Like all other crustaceans, shrimps have a hard exoskeleton or shell. Their shells are made from a substance called chitin, which is a very strong and hard protein. In this experiment, you are going to use an acid (vinegar), a base (bleach), and a control (salt water) to see if you can to dissolve the shrimp shell.

Hypothesis

Can salt water dissolve shrimp shells?   Yes   No

Can bleach dissolve shrimp shells?   Yes   No

Can vinegar dissolve shrimp shells?   Yes   No

Materials

_____   _____

_____   _____

_____   _____

_____   _____

_____   _____

Procedure

_____

_____

_____

_____

_____

_____

_____

_____

_____

_____

Observations and Results

_____
_____
_____
_____
_____
_____
_____
_____

Conclusion

_____
_____
_____
_____
_____
_____

Written Assignment Week 15

Discussion Questions

1. Name several of the animals that are a part of the crustacean class.
2. What are two characteristics of crustaceans?
3. How do crustaceans begin life?
4. What is molting?
5. What are two ways that crabs use their pincers?

_____

_____

_____

_____

_____

_____

_____

_____

_____

_____

_____

_____

_____

_____

_____

_____

_____

_____

_____

_____

_____

_____

Written Assignment Week 15

Student Assignment Sheet Week 16
Insects (Arthropods)

Experiment: Do insects play a part in decomposition?
Materials
- ✓ 1 Apple
- ✓ 2 Jars, one that has a lid

Procedure
1. Read the introduction to this experiment and answer the question.
2. Cut the apple in half, place one half in a jar with no lid, place the other half in the other jar, and cover with a tightly fitting lid.
3. Set both jars outside in a place where they won't be disturbed. Check the jars daily and write your observations on the experiment sheet.
4. After 12 days, read over your daily observations and draw conclusions on the experiment sheet.

Vocabulary & Memory Work
- ☐ Vocabulary: antenna, life cycle, metamorphosis, nymph, thorax
- ☐ Memory Work—Continue to work on memorizing the Basic Phyla of the Animal Kingdom and their characteristics. This week add:

   9. Chordates – Animals whose bodies are supported by a stiff rod called a notochord (*Example: Vertebrates*).

Sketch: Life cycle of a butterfly
- 🖼 Label with the following – female lays eggs on a leaf; caterpillars hatch and eat the leaf; mature caterpillar spins thread, sheds skin and forms chrysalis; inside the completed chrysalis, the adult butterfly forms; the chrysalis skin splits and the adult butterfly emerges.

Writing
- ∽ Reading Assignment: *Kingfisher Science Encyclopedia of Nature* pp. 75-77 Insects
- ∽ Additional Research Reading
   - 📖 Insects: *DKEN* pp. 162-163
   - 📖 Bugs: *DKEN* pp. 170-171

Dates to Enter
- 🕐 3000 BC – Scarab beetles held religious and cultural symbolism in Old Egypt.

Sketch Assignment Week 16

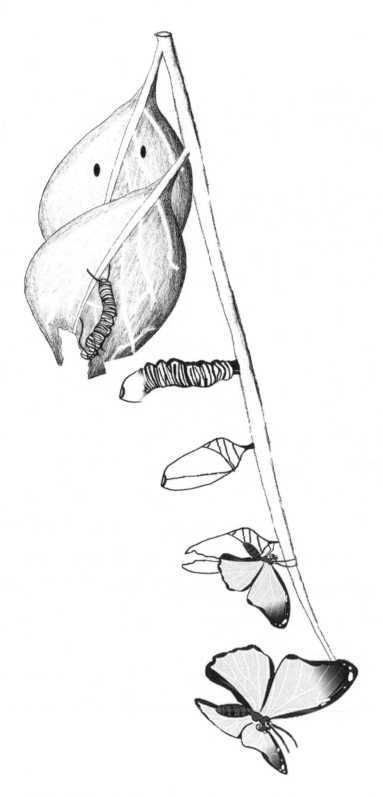

116

Experiment: Do insects play a part in decomposition?

Introduction

Decomposition is the breaking down of organic matter. Decomposers are living things that obtain food by breaking down the remains of other living things. Fungi are decomposers, as are many types of bacteria. In this experiment, you will test to see if insects are also decomposers and see whether the presence of insects speeds up or slows down decomposition.

Hypothesis

Do insects play a part in decomposition?   Yes   No

Materials

_____   _____

_____   _____

_____   _____

_____   _____

_____

Procedure

_____

_____

_____

_____

_____

_____

_____

_____

_____

_____

_____

Observations

Day 1 – _____

Day 2 – _____

Day 3 – _____

Day 4 – _____

Day 5 – _____

Day 6 – _____

Day 7 – _____

Day 8 – _____

Day 9 – _____

Day 10 – _____

Day 11 – _____

Day 12 – _____

Conclusion

_____

_____

_____

Written Assignment Week 16

Discussion Questions

1. Where do insects live?
2. What are the 3 sections of the insect's body and what is found in each section?
3. What does it mean to be a social insect?
4. How many pairs of wings do most insects have and what do they use them for?
5. What are the two life cycles that insects can go through? (Explain both.)

Written Assignment Week 16

Biology: Unit 4

Vertebrates

Biology Unit 4: Vertebrates
Vocabulary Sheet

Define the following terms as they are assigned on the Student Assignment Sheet.

1. Gills – _____

    _____

    _____

2. Swim bladder – _____

    _____

    _____

3. Vertebrate – _____

    _____

    _____

4. Amphibian – _____

    _____

    _____

5. Tadpole – _____

    _____

    _____

6. Ectotherm – _____

    _____

    _____

7. Reptile – _____

    _____

    _____

8. Scales – _____

    _____

    _____

9. Clutch – _____

_____

_____

10. Crop – _____

_____

_____

11. Gizzard – _____

_____

_____

12. Fledgling – _____

_____

_____

13. Endotherm – _____

_____

_____

14. Mammals – _____

_____

_____

15. Mammary gland – _____

_____

_____

16. Monotreme – _____

_____

_____

Student Assignment Sheet Week 17
Fish

Experiment: Which fish will float higher?

Materials
- ✓ 1 large clear glass jar or bowl
- ✓ 3 small balloons
- ✓ 3 small marbles
- ✓ Ruler
- ✓ Water

Procedure
1. Read the introduction to this experiment and answer the question.
2. Fill the glass cup or bowl with cool water. Label each balloon with #1, #2, or #3. Stretch each balloon and then place a marble in each.
3. Tie balloon #1 off with just the marble in it and no air. Blow up balloon #2 a little bit and tie it off. Blow up balloon #3 about halfway and tie it off.
4. Place each balloon the water and let them come to rest (about one minute). Then measure where each one stays.
5. Record the measurement and be sure to include the units. Then, complete the experiment sheet.

Vocabulary & Memory Work

- ☐ Vocabulary: gills, swim bladder, vertebrate
- ☐ Memory Work—Begin working on the Basic Classes of the Phyla Chordata. This week work on:
  - **1. Fish** – Cold-blooded animals that live in water; they are covered with scales and breathe through gills.

Sketch: Anatomy of a fish

- ▨ Label the following – mouth, eye, brain, spinal cord, lateral line, swim bladder, heart, gills, dorsal fin, caudal fin, anal fin, pelvic fin, pectoral fin

Writing

- ⤸ Reading Assignment: *Kingfisher Science Encyclopedia* pp. 78-79 Fish
- ⤸ Additional Research Reading
  - 📖 Fish: *DKEN* pp. 184-185
  - 📖 Movement in Water: *USE* pp. 304-305, *DKEN* pp. 38-39

Dates to Enter

- 🕐 2003 – Scottish scientists at the University of Edinburgh and the Roslin Institute concluded that rainbow trout exhibit behaviors often associated with pain in other animals.

Sketch Assignment Week 17

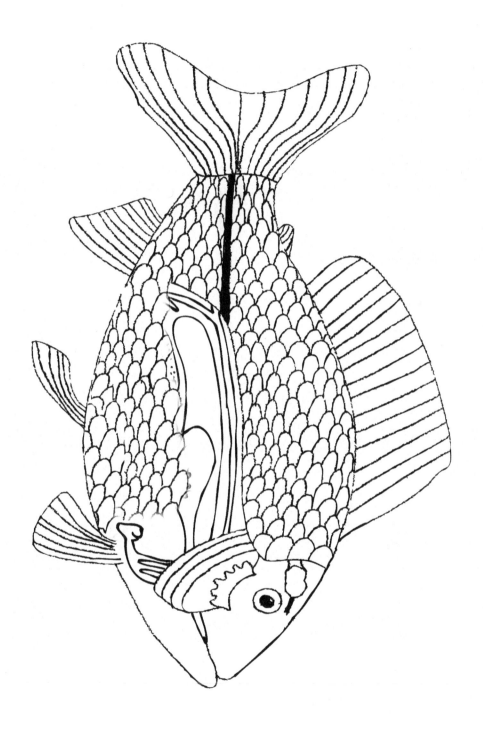

Experiment: Which fish will float higher?

Introduction

Fish are a part of the phyla chordata, which means that they have a backbone. They also have a unique organ called a swim bladder. The swim bladder is an organ that helps to control the fish's buoyancy. Buoyancy is the ability to float in a liquid. In this experiment, we are going to look at how the amount of air in the swim bladder (or balloon in our case) affects the buoyancy

Hypothesis

I believe that the balloon with the _____ amount of air in it will float the
(most, a little or no)

highest.

Materials

_____   _____

_____   _____

_____   _____

_____   _____

_____   _____

Procedure

_____

_____

_____

_____

_____

_____

_____

_____

_____

_____

_____

_____

Observations

Balloon #1 floated _____ high

Balloon #2 floated _____ high

Balloon #3 floated _____ high

_____

_____

_____

_____

Conclusion

_____

_____

_____

_____

_____

_____

Written Assignment Week 17

Discussion Questions

1. What are the three types of fish and their main characteristics?
2. How do saltwater and freshwater fish get rid of excess water from their bodies?
3. How do fish usually reproduce?
4. What is the lateral line and how does a fish use it?
5. Explain how a fish breathes in the water.

Written Assignment Week 17

Student Assignment Sheet Week 19
Amphibians

Experiment: Frog dissection

    Materials

      ✓ 1 Frog dissection kit

      ✓ 1 Preserved frog

    Procedure

      1. Please follow all the directions included in the dissection kit. There is no experiment write-up for this week.

Vocabulary & Memory Work

    ☐ Vocabulary: amphibian, tadpole

    ☐ Memory Work—Continue to work on the Basic Classes of the Phyla Chordata. This week add the following:

      2. Amphibians – Cold-blooded animals that live on land and in the water; they have soft skin.

Sketch: Life cycle of a frog

    ▨ Label the following – embryos develop in eggs laid by mature adults, tadpole hatches from the egg, tadpoles grows, tadpole develops legs, young moves from the water onto the land, adult frog matures

Writing

    ↪ Reading Assignment: *Kingfisher Science Encyclopedia* pp. 80-81 Amphibians

    ↪ Additional Research Reading

      📖 Amphibians: *DKEN* pp. 192-193

      📖 Life Cycles: *USE* pp. 328-329

Dates to Enter

    🕐 2008 – The Evolutionarily Distinct and Globally Endangered (EDGE) identified nature's most endangered species, eight of which are amphibians.

Sketch Assignment Week 18

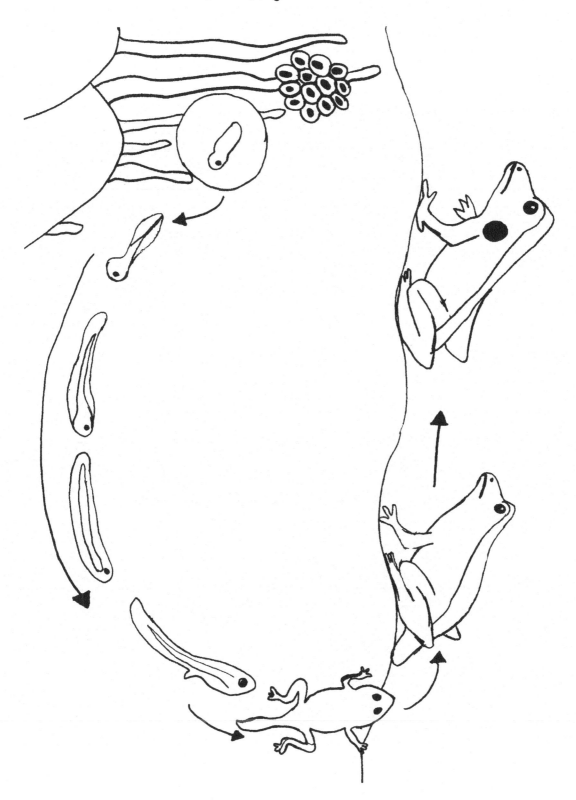

Written Assignment Week 18

Discussion Questions

1. What are three characteristics of amphibians?
2. What are the three main groups of amphibians?
3. How can you tell the difference between frogs and toads?
4. How do some amphibians, like the poison dart frog, avoid predators?
5. Briefly describe the typical amphibian life cycle.

Written Assignment Week 18

Student Assignment Sheet Week 19
Reptiles

Experiment: Do I use only my tongue to taste things?
Materials
- ✓ Clothespin
- ✓ Blindfold
- ✓ 5 Plates
- ✓ 5 Pieces of bread with different edible spreads on them (such as garlic, cinnamon/ sugar, butter, peanut butter, and plain)

Procedure
1. Read the introduction to this experiment and answer the question.
2. Prepare the bread with the various edible spreads on five different plates and number the samples from #1-5.
3. Use the clothespin to close your nose. Then have your partner blindfold you and mix up the plates.
4. Have your partner feed each food to you, while you try to guess what each one is.
5. Remove the clothespin. Then taste each of the samples again.
6. Record the results and complete the experiment sheet.

Vocabulary & Memory Work
- ☐ Vocabulary: ectotherm, reptile, scales
- ☐ Memory Work—Continue to work on the Basic Classes of the Phyla Chordata. This week add the following:
  3. Reptiles – Cold-blooded animals that lay eggs; they are covered with scales.

Sketch: Chameleon Defense
🖾 A chameleon's defense is to change its color. Color the first chameleon pale green with yellow stripes; label it "normal chameleon." Color the second chameleon dark brown with light brown stripes; label it "chameleon when threatened."

Writing
- ⤾ Reading Assignment: *Kingfisher Science Encyclopedia* pp. 82-83 Reptiles
- ⤾ Additional Research Reading
  - 📖 Reptiles: *DKEN* pp. 198-199
  - 📖 Body Covering: *USE* pp. 302-303

Dates to Enter
- 🕓 1866 – Haeckel demonstrated that vertebrates could be divided based on their reproductive strategies, and that reptiles, birds, and mammals were united by the amniotic egg. By the end of the 19th century, the class *Reptilia* had come to include all the amniotes except birds and mammals.

Sketch Assignment Week 19

Experiment: Do I use only my tongue to taste things?

Introduction

Reptiles use their sense of sight, smell, and hearing to find their food. Snakes and some lizards have a special sense organ on the roof of their mouth called the Jacobson's organ. This organ gives them the ability to taste and smell chemical particles in the air with their tongue. In this experiment, you will test to see if we use only our tongues to taste our food.

Hypothesis

Do I use only my tongue to taste things?　　　Yes　　　No

Materials

_____　　　_____
_____　　　_____
_____　　　_____
_____　　　_____
_____　　　_____

Procedure

_____
_____
_____
_____
_____
_____
_____
_____
_____
_____
_____
_____

Observations & Results

	Sample	My guess with the clothespin & blindfold on	My guess with the clothespin & blindfold off
1			
2			
3			
4			
5			

Conclusion

_____

_____

_____

_____

_____

_____

_____

_____

Written Assignment Week 19

Discussion Questions

1. What are some characteristics of a reptile?
2. What are the four main groups of reptiles?
3. How do reptiles control their body temperature?
4. What are some characteristics of crocodilians?
5. What characteristic is unique to tortoises and turtles?
6. What are some characteristics of snakes?
7. What do some snakes and lizards do that most reptiles do not?

Written Assignment Week 19

Student Assignment Sheet Week 20
Birds

Experiment: Which beak is best suited for picking up which type of food?

Materials

- ✓ Chopsticks
- ✓ Tweezers
- ✓ Pliers
- ✓ Eye dropper
- ✓ Sugar water or honey
- ✓ Gummy worms
- ✓ Unshelled peanuts
- ✓ Seeds
- ✓ Raisins
- ✓ Plates

Procedure

1. Read the introduction to this experiment and match the type of beak to the type of food you think its best suited for picking up.
2. Prepare five plates; one for each of the different types of food. Determine the scale you would like to use when rating ease of pickup. For example, the scale range could be from 1 to 10, with "1" standing for the easiest and "10" standing for the hardest. Fill in the scale range on the experiment sheet.
3. Use the chopsticks to try to pick up each of the different types of food. Rate the ease of pickup on the chart on the experiment sheet. Repeat using the tweezers, pliers, and eye dropper.
4. Draw conclusions and complete the experiment sheet.

Vocabulary & Memory Work

- ☐ Vocabulary: clutch, crop, gizzard, fledgling
- ☐ Memory Work—Continue to work on the Basic Classes of the Phyla Chordata. This week add the following:
  4. Birds – Warm-blooded animals that lay eggs; they have feathers and wings.

Sketch: Three Types of Bird Feet

- 🖾 Label the following: hooked claws and talons from a bird of prey, a backward-pointing toe from a perching bird, webbed feet from a swimming bird

Writing

- ᕦ Reading Assignment: *Kingfisher Science Encyclopedia* pp. 84-85 Birds
- ᕦ Additional Research Reading
  - 📖 Birds: *DKEN* pp. 208-209
  - 📖 Movement in Air: *USE* pp. 306-307, *DKEN* pp. 36-37

Dates to Enter

- ⏲ 1676 – The first classification of birds was developed by Francis Willoughby and John Ray in their volume *Ornithologiae*.
- ⏲ 2003 – Domesticated birds raised for meat and eggs, called poultry, are the largest source of animal protein eaten by humans.

Sketch Assignment Week 20

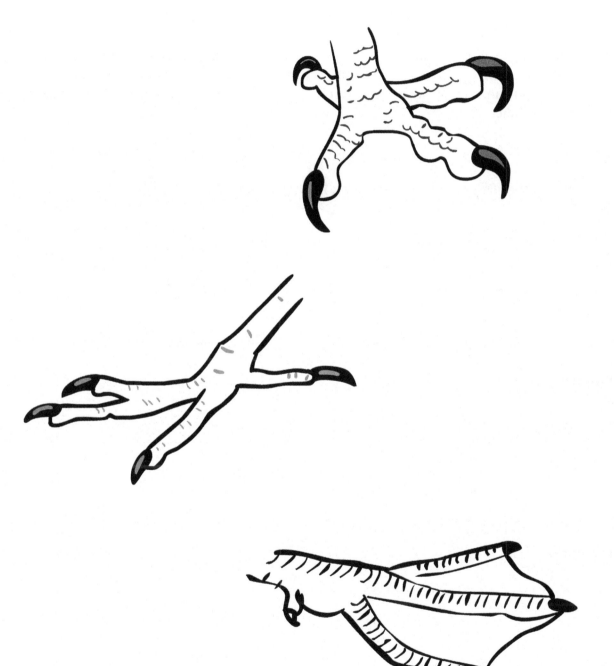

Experiment: Which beak is best suited for picking up which type of food?

Introduction

All birds have bills or beaks that vary in shape and size. Birds use their beaks to obtain and manipulate their food. They also use their beaks for preening their feathers and for building their nests. Some species of parrots even use their beaks for climbing. In this experiment, you are going to test to see how the size and shape of a beak relates to the bird's diet.

Hypothesis

Match the beak to the food you think it will best be able to pick up:

Long, large beak
(Chopsticks)

sugar water or honey

Short, light beak
(Tweezers)

unshelled peanuts

Short, strong beak
(Pliers)

bread   seeds

Long, slender beak
(Eye Dropper)

worms   raisins

Materials

_____   _____

_____   _____

_____   _____

_____   _____

_____   _____

Procedurex

_____

_____

_____

_____

_____

_____

_____

_____

_____

_____

_____

Observations and Results

Rate how easy the food was to pick up using the scale you determine.

Scale Range: _____

Easiest = _____ Hardest = _____

	Bread	Gummy Worms	Unshelled Peanuts	Sugar Water	Seeds	Raisins
Chopsticks						
Tweezers						
Pliers						
Eye Dropper						

Conclusion

_____

_____

_____

_____

_____

_____

_____

_____

Written Assignment Week 20

Discussion Questions

1. What are three characteristics that enable birds to be so efficient at flight?
2. What do birds use their feathers for?
3. Why do birds regularly preen?
4. What is the syrinx and what is its purpose?
5. How do all birds reproduce?
6. Name two adaptations that birds have for flight.

Written Assignment Week 20

Student Assignment Sheet Week 21
Mammals

Experiment: Which type of fur keeps mammals warmer?

Materials
- ✓ Felt
- ✓ Cotton balls
- ✓ Water
- ✓ 1 Large cup
- ✓ 4 Small cups
- ✓ Instant read thermometer
- ✓ Foil

Procedure
1. Read the introduction to this experiment and circle the type of "fur" you think will hold in the most warmth.
2. Line the outside of cup #1 with felt and fill three-quarters of the way with water and label it.
3. Next, take the large cup and place a few cotton balls at the bottom of the cup. Then place cup #2 on top of the cotton balls, and fill in the space with more cotton balls. Then fill cup #2 three-quarters of the way with water and label it.
4. Glue a layer of cotton ball onto a layer of felt and wrap it around cup #3, and then cover it with another layer of felt. Fill cup #3 three-quarters of the way with water and label it.
5. Fill cup #4 three-quarters of the way with water and label it. This is the control.
6. Take the initial temperature of all four cups, record, and cover with foil. Place all four cups into the freezer.
7. Measure and record the temperatures every thirty minutes over a period of three hours.
8. Draw conclusions and complete the experiment sheet.

Vocabulary & Memory Work
- ☐ Vocabulary: endotherm, mammals, mammary gland, monotreme
- ☐ Memory Work—Continue to work on the Basic Classes of the Phyla Chordata. This week add the following:

  5. Mammals – Warm-blooded animals that feed their young milk; they are covered with fur.

Sketch: Mitosis
- 🖾 Label the following – chromosomes change into thick coils, chromosomes line up in the center, the chromosomes pairs split apart, cells pinch apart, daughter cells grows.

Writing
- ḇ Reading Assignment: *Kingfisher Science Encyclopedia* pp. 86-87 Mammals, pp. 88-89 Animal Reproduction
- ḇ Additional Research Reading
  - 📖 Mammals: *DKEN* pp. 232-233
  - 📖 Movement on Land: *USE* pp. 308-309, *DKEN* pp. 34-35
  - 📖 Mitosis (cell division): *UIDS* pp. 240-241

Dates to Enter
- 🕐 No dates to enter this week.

Sketch Assignment Week 21

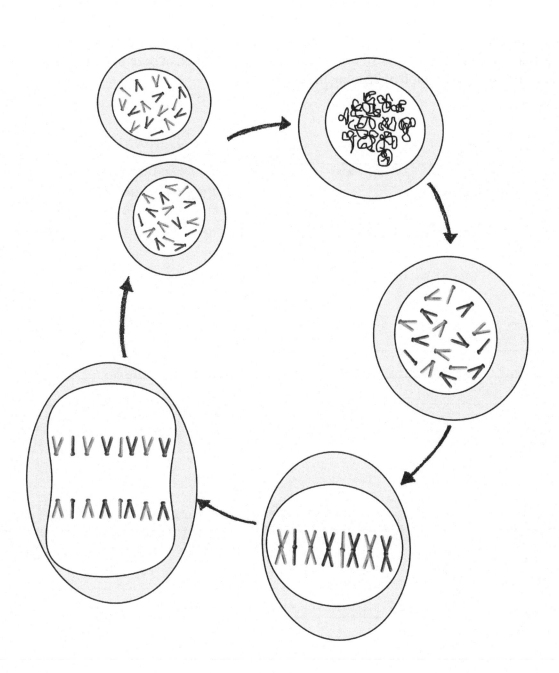

Experiment: Which type of fur keeps mammals warmer?

Introduction

All mammals are warm-blooded, which means that they are able to maintain a constant internal body temperature no matter what the outside temperature is. Mammals use their food to make energy that helps keep them warm. They also have fur to serve as a barrier to cold air and water. In this experiment, we will look at several types of fur to see which does the best job of keeping the cold out.

Hypothesis

I think that _____ will keep a mammal the warmest.

Thin fur (felt only)   Thick fur (cotton balls)

Very Thick fur (felt & cotton balls)   No fur

Materials

_____   _____

_____   _____

_____   _____

_____   _____

_____   _____

Procedure

_____

_____

_____

_____

_____

_____

_____

_____

_____

_____

Observations and Results

Temperature	Initial	After 30 min	After 60 min	After 90 min	After 120 min	After 150 min	After 180 min
Cup #1							
Cup #2							
Cup #3							
Cup #4							

Conclusion

_____

_____

_____

_____

_____

_____

Written Assignment Week 21

Discussion Questions

Mammals, pp. 86-87

1. What are two other characteristics of most mammals?
2. What are the three groups of mammals?
3. Explain how the three groups of mammals differ in the way that they reproduce.
4. What two factors have contributed to the success of mammals?

Animal Reproduction, pp. 88-89

1. How is animal's lifespan related to growth and reproduction?
2. Which group of chordates takes the longest to reproduce? Why?
3. What are clones?

Written Assignment Week 21

Biology
Unit 5

Animal Overview

Biology Unit 5 Animal Overview
Vocabulary Sheet

Define the following terms as they are assigned on the Student Assignment Sheet.

1. Hemisphere – _____

   _____

   _____

2. Migration – _____

   _____

   _____

3. Camouflage – _____

   _____

   _____

4. Predator – _____

   _____

   _____

5. Prey – _____

   _____

   _____

6. Autotroph – _____

   _____

   _____

7. Heterotroph – _____

   _____

   _____

8. Carnivore – _____

   _____

   _____

9. Herbivore – _____

_____

_____

10. Omnivore – _____

_____

_____

11. Symbiosis – _____

_____

_____

Student Assignment Sheet Week 22
Animal Research Project

Science Fair Project

This week, you will complete step one and begin step two of your Science Fair Project. You will be choosing your topic, formulating a question and doing some research about that topic.

1. **Choose your topic** – You should choose a topic in the field of biology that interests you, such as plants. Next, come up with several questions you have relating to that topic, (e.g. "Why do plants grow?" or "How fast do plants grow?"). Then, choose the one question you would like to answer and refine it (e.g. "Does the amount of light a plant gets effect how quickly it grows?").

2. **Do Some Research** – Now that you have a topic and a question for your project, it is time to learn more about your topic so that you can make an educated guess (hypothesis) on the answer to your question. For the question stated above, you would need to research topics like the structure plant and the factors that effect its growth. Begin by looking up the topic in the references you have at home. Then, make a trip to the library to search for more on the topic. As you do your research, write any relevant facts you have learned on index cards and be sure to record the sources you use.

Animal Research Project: Creating a Profile Page

This week, you will spend part of the week on researching an animal (either invertebrate or vertebrate) and creating a profile page for that animal. Follow the steps below:

1. **Choose an animal** – Choose an animal, vertebrate or invertebrate for an in-depth profile. You can look at the animals picture in *Kingfisher Science Encyclopedia* pp. 70-87 for inspiration.

2. **Do some research** – Answer the following questions about the animal:
   - ✓ What is the scientific name of the animal?
   - ✓ What is the normal size range for the animal?
   - ✓ What habitat does the animal live in?
   - ✓ Where is the animal typically found (i.e. what country)?
   - ✓ What does the animal typically eat?
   - ✓ What are some interesting facts or characteristics of the animal?

   Read about your chosen animal on-line, at the library, or in an encyclopedia and answer the questions that you can. Write down any interesting facts you have learned on individual index cards.

3. **Complete the profile page for the animal** – Write the common name for the animal on the blank at the top of the page. Then fill out the remaining information. The brief description should be four to five sentences long. It should include some of the interesting facts you learned about the animal from your research. Then draw the animal or glue a printed picture of it in the box.

Vocabulary & Memory Work

- ☐ Vocabulary: No vocabulary this week
- ☐ Memory Work—Begin working on Animal Defenses. (*See the Student Guide Appendix pp. 248-249 for a complete list.*)

Student Guide Biology Unit 5: Animal Overview ~ Week 22 Animal Research Project

Animal Profile Page

_____

Scientific Name

_____

Size

_____

Habitat

_____

Location

_____

Diet

_____

_____

Brief Description

_____

_____

_____

_____

_____

_____

_____

_____

_____

_____

_____

Science Fair Project Step 1: Choose a Topic

Key 1 ~ Decide on an area of science.

What areas of biology are you interested in learning about?

_____

_____

_____

_____

Rank your interest in the different areas you listed and then circle the one area that you would like to use for your topic.

Key 2 ~ Develop several questions about the area of biology.

What questions would you like to answer about your area of biology? (***Note:*** *Remember that good questions begin with how, what, when, who, which, why or where.*)

_____

_____

_____

_____

_____

_____

Key 3 ~ Choose a question to be the topic.

Write down the question that you will be using for your project.

_____

_____

_____

Science Fair Project Step 2: Do Some Research

Key 1 ~ Brainstorm for research categories.

What categories are you going to research for your project?

1. _____

2. _____

3. _____

4. _____

5. _____

Key 2 ~ Research the categories.

Use the following template for your research cards:

Category Number	Reference Letter
One piece of Information	

Record your sources below.

A. _____

_____

B. _____

_____

C. _____

_____

D. _____

_____

E. _____

_____

F. _____

_____

Student Assignment Sheet Week 23
Migration

Science Fair Project

This week, you will complete steps two through four of your Science Fair Project. You will be finishing your research, formulating your hypothesis and designing your experiment.

2. **Do Some Research** – This week, you will finish your research. Then, organize your research index cards and write a brief report on what you have found out.

3. **Formulate a Hypothesis** – A hypothesis is an educated guess. For this step, you need to review your research and make an educated guess about the answer to your question. A hypothesis for the question asked in step one would be, "The more light there is, the better the plant will grow."

4. **Design an Experiment** – Your experiment will test the answer to your question. It needs to have a control and several test groups. Your control will have nothing changed, while your test groups will change only one factor at a time. An experiment to test the hypothesis given above would be to grow three pots of grass on a shelf in your room; measure and record the growth daily on a graph you design. After five days, move one plant into a closet with no light, move one plant onto a window sill in the sun, and leave the other plant where it is. If time allows, you can go ahead and begin your experiment this week.

Vocabulary & Memory Work

- [ ] Vocabulary: hemisphere, migration
- [ ] Memory Work—Continue to work on Animal Defenses.

Sketch: Migration Routes

- Trace the routes of the swallow, reindeer, monarch butterfly, arctic tern, and humpback whale. Do each route in a different color and include a key for them.

Writing

- Reading Assignment: *Kingfisher Science Encyclopedia* pg. 93 Migration
- Additional Research Reading
  - Migration and Navigation: *DKEN* pp. 46-47
  - Migration (section): *USE* pg. 329

Dates to Enter

- No dates to be entered this week.

Sketch Assignment Week 23

Science Fair Project Step 2: Do Some Research

Key 3 ~ Organize the information.

Organize the information for your report.

Key 4 ~ Write a brief report.

Write down what the order of your categories will be for your report.

1. _____

2. _____

3. _____

4. _____

5. _____

On a separate sheet of paper write out a rough draft of your research report.

Science Fair Project Step 3: Formulate a Hypothesis

Key 1 ~ Review the Research.

Read over your research.

Key 2 ~ Formulate an Answer.

Write down your hypothesis for your science fair project.

_____

_____

_____

_____

_____

Science Fair Project Step 4: Design an Experiment

Key 1 ~ Choose a Test.

What are some ways that you can test your hypothesis?

_____

_____

_____

_____
_____
_____

Key 2 ~ Determine the Variables.

What factor are we trying to test? (Independent variable)

_____

What factor will we use to measure the progress of our test? (Dependent variable)

_____

What factors do we need to keep constant so that they will not affect our results? (Controlled variables)

_____
_____

Key 3 ~ Plan the Experiment.

What will the groups in your experiment be?

Control Group: _____

Test Group 1: _____

Test Group 2: _____

Test Group 3: _____

Test Group 4: _____

Write down the plan for your experiment.

_____
_____
_____
_____
_____
_____
_____

Written Assignment Week 23

Discussion Questions

1. Why do animal migrate?
2. How do animals find their way during migration?
3. Which animals are the most common migrants?
4. Name two ways that birds use to navigate.

_____

_____

_____

_____

_____

_____

_____

_____

_____

_____

_____

_____

_____

_____

_____

_____

_____

_____

_____

_____

_____

_____

Written Assignment Week 23

Student Assignment Sheet Week 24
Animal Defenses

Science Fair Project

This week, you will complete steps five and six of your Science Fair Project. You will carry out the experiment and record your observations and results.

5. **Perform the Experiment** – This week, you will perform the experiment you designed last week. Be sure to take pictures along the way as well as record your observations and results. (**Note** – *Observations are a record of the things you see happening in your experiment. For instance, an observation would be that the plant in the light is green and tall, but the plant in the dark is turning yellow and has not grown in days. Results are specific and measurable. For instance, results would be that the plant in the light has grown 2 inches over 5 days and the plant in the dark has grown 0 inches in 5 days. Observations are generally recorded in journal form, while results can be compiled into tables, charts, and graphs or relayed in paragraph form.*)

6. **Analyze the Data** – Once you have compiled your observations and results, you can use them to answer your question. You need to look for trends in your data and make conclusions from that. A possible conclusion to the electrolysis experiment would be, "Grass needs light to grow. The more light that grass is exposed to the better it will grow." If your hypothesis does not match your conclusion or your were not able to answer your question using the results from your experiment, you may need to go back and do some additional experimentation.

Vocabulary & Memory Work
- Vocabulary: camouflage, predator, prey
- Memory Work—Continue to work on Animal Defenses.

Sketch
- There is no sketch for this week to allow more time to work on the science fair project.

Writing
- Reading Assignment: *Kingfisher Science Encyclopedia* pg. 95 Adaptation and Defense
- Additional Research Reading
  - Defense: *DKEN* pp. 48-51
  - Sending Messages: *USE* pp. 318-319

Dates to Enter
- No dates to be entered this week.

Science Fair Project Step 5: Perform the Experiment

Key 1 ~ Get ready for the experiment.

When do you plan to run your experiment?

From _____ to _____.

Purchase and gather your materials, and prepare any of the materials that need to be pre-made.

Key 2 ~ Run the experiment.

What things do you need to remember to do each day?

Take pictures of the experiment every day or for every trial.

_____

_____

_____

_____

_____

_____

_____

_____

_____

Key 3 ~ Record any Observations and Results

Record your observations and results on a separate sheet of paper.

Science Fair Project Step 6: Analyze the Data

Key 1 ~ Review and organize the data.

What trends did you recognize in your observations?

_____

_____

_____

_____

_____

What information did you interpret from your results?

_____

_____

_____

_____

_____

_____

_____

Key 2 ~ State the answer.

After reviewing your data, write the answer to your question. (**Note**—*Your statement should begin with "I found that..." or "I discovered that..."*)

_____

_____

_____

_____

_____

Key 3 ~ Draw several conclusions.

Answer the following questions:

- ✓ Was my hypothesis proven true? (**Note**—*If your hypothesis was proven false, be sure to state why you think it was proven false.*)
- ✓ Did you have any problems or difficulties when performing your experiment?
- ✓ Did anything interesting happen that you would like to share?
- ✓ Can you think of any other things related to your project that you would like to test in the future?

Now take your answer from key two and your answers from key three to write your conclusion on a separate sheet of paper. Your paragraph should be four to six sentences in length.

Science Fair Project Observations and Results

Observations

_____

_____

_____

_____

_____

_____

_____

_____

_____

Results Chart:

Time	Control Group	Test Group 1	Test Group 2	Test Group 3	Test Group 4

Written Assignment Week 24

Discussion Questions

1. What is the difference between adaptation and defense?
2. What are the five main types of defensive weapons (give examples of each)?
3. What are the two different types of adaptations?
4. What are genotypic adaptations? Phenotypic adaptations?

Written Assignment Week 24

Student Assignment Sheet Week 25
Feeding and Nutrition

Science Fair Project

This week, you will complete steps seven and eight of your Science Fair Project. You will be writing and preparing a presentation of your Science Fair Project.

7. **Create a Board** – This week, you will be creating a visual representation of your science fair project that will serve as the centerpiece of your presentation. You will begin by planning the look of your board, then move onto preparing the information and finally you will pull it all together.

8. **Give a Presentation** – After you have completed your presentation board, determine if you would like to include part of your experiment in your presentation. Then, prepare a 5 minute talk about your project, be sure to include the question you tried to answer, your hypothesis, a brief explanation of your experiment, and the results plus the conclusion to your project. Be sure to arrive on time for your presentation. Set up your project board and any other additional materials. Give your talk and then ask if there are any questions. Answer the questions and end your time by thanking whoever has come to listen to your presentation.

Vocabulary & Memory Work

☐ Vocabulary: autotroph, heterotroph, carnivore, herbivore, omnivore, symbiosis
☐ Memory Work—Continue to work on Animal Defenses.

Sketch

☒ There is no sketch for this week to allow more time to work on the science fair project.

Writing

☞ Reading Assignment: *Usborne Science Encyclopedia* pp. 310-311 Feeding, pp. 312-313 Teeth and Digestion
☞ Additional Research Reading
   📖 Animal Feeding: UIDS pp. 270-271
   📖 Feeding and Nutrition: DKEN pp. 22-23

Dates to Enter

🕐 No dates to be entered this week.

Science Fair Project Step 7: Create a Board

Key 1 ~ Plan out the board.

Use the template on the following page to sketch out a rough plan for your presentation board.

Key 2 ~ Prepare the information.

Type up the following information for your presentation board:

- ☐ Introduction
- ☐ Hypothesis
- ☐ Research
- ☐ Materials
- ☐ Procedure
- ☐ Results
- ☐ Conclusion

Key 3 ~ Put the board together.

- ☐ Put the decorative elements on your project board.

- ☐ Print out and attach your information paragraphs.

- ☐ Add the title to your science fair project board.

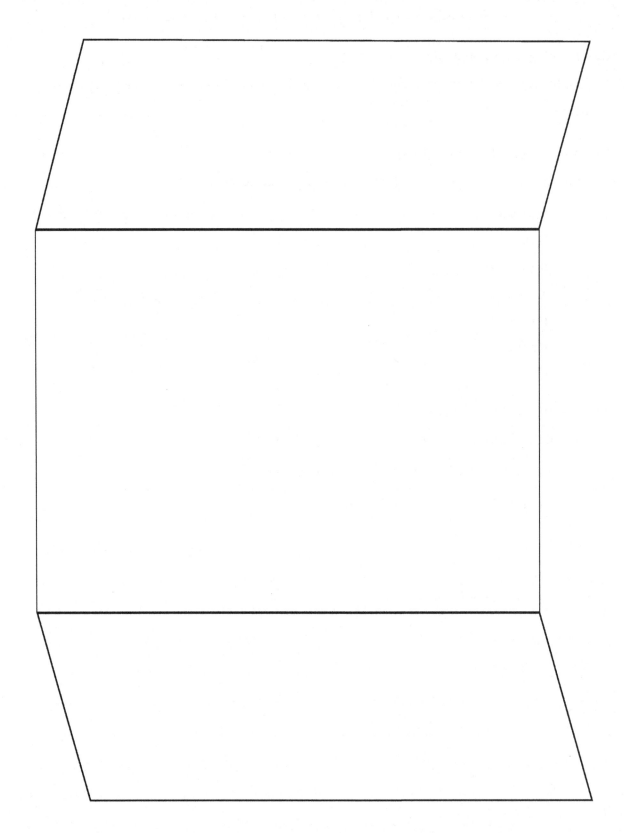

Student Guide Biology Unit 5: Animal Overview ~ Week 25 Feeding & Nutrition

Science Fair Project Step 8: Give a Presentation

Key 1 ~ Prepare the presentation.

Write down the outline for your presentation on a separate sheet of paper.

Key 2 ~ Practice the presentation.

☐ Practice your presentation in front of a mirror several times.

☐ Practice your presentation with your teacher.

Key 3 ~ Share the presentation.

Keep the following tips in mind for your presentation:

- ✓ Arrive on time for your presentation.
- ✓ Set up your project board and any other additional materials.
- ✓ Give your talk and then ask if there are any questions.
- ✓ Answer the questions and end your time by thanking whomever has come to listen to your presentation.

Written Assignment Week 25

Discussion Questions

Feeding, pp. 310-311

1. What does the structure of an animal's mouthparts depend upon?
2. What is phagocytosis?
3. How do most animals that live in the water feed?
4. What are the different parts of an insect's mouth and what are they used for?

Teeth and Digestion, pp. 312-313

1. How do a carnivore's teeth differ from an herbivore's teeth?
2. What can omnivores eat?
3. What two adaptations do birds have in their digestive system to deal with solid food?

Written Assignment Week 25

Biology:
Unit 6

The Human Body

Biology Unit 6: The Human Body, part 1
Vocabulary Sheet

Define the following terms as they are assigned on the Student Assignment Sheet.

1. Epidermis – _____

   _____

   _____

2. Dermis – _____

   _____

   _____

3. Cuticle – _____

   _____

   _____

4. Cartilage – _____

   _____

   _____

5. Axial skeleton – _____

   _____

   _____

6. Appendicular skeleton – _____

   _____

   _____

7. Myofibrils – _____

   _____

   _____

8. Involuntary muscle – _____

   _____

   _____

9. Voluntary muscle – _____

_____

_____

10. Neuron – _____

_____

_____

11. Sensory neuron – _____

_____

_____

12. Motor neuron – _____

_____

_____

13. Hormone – _____

_____

_____

14. Endocrine gland – _____

_____

_____

Student Assignment Sheet Week 26
Integumentary System (Skin, Hair, and Nails)

Experiment: Which is stronger, regular hair or bleached hair?
Materials
- ✓ Pennies (10-30)
- ✓ 2 Pieces of hair (at least 5 inches long)
- ✓ Several heavy books
- ✓ Pencil, Tape, Glass, Bleach, Gloves, Water

Procedure
1. Read the introduction to this experiment and answer the question.
2. Place one of the hairs in the glass and add in ¼ cup (60 mL) of water plus 1 TBSP (15 mL) of bleach into the glass. Let it sit for fifteen minutes.
3. Then, make a stack of the heavy books that is at least 10 inches high.
4. Tie one end of the other hair to one end of the pencil, and then put the other end of that pencil into the books so that it is held firmly in place.
5. Tape one penny onto the other end of the hair. Continue adding pennies until the hair strand breaks. Record how many pennies on the experiment sheet.
6. Put on the gloves and remove the hair strand from the glass with the bleach. Rinse it thoroughly.
7. Repeat Steps 4 and 5 with the bleached hair strand.
8. Then, complete the experiment sheet.

Vocabulary & Memory Work
- ☐ Vocabulary: epidermis, dermis, cuticle
- ☐ Memory Work—Begin working on the Body Systems
  - ✓ **Integumentary System** – It covers & protects the body.

Sketch: Anatomy of Skin
- ▨ Label the following – hair shaft, epidermis, dermis, sweat gland, oil gland, sweat pore, hair root, fat, blood vessels, nerve, muscle

Writing
- ⁀ Reading Assignment: *Kingfisher Science Encyclopedia* pp. 100-101 Skin, Hair, Nails
- ⁀ Additional Research Reading
  - 📖 Skin, Hair, Nails: *USE* pp. 368-369
  - 📖 The Skin: *UIDS* pp. 310-311
  - 📖 Skin & Touch: *DK HB* pp. 32-33

Dates to Enter
- 🕑 c1500BC – The earliest known medical text Ebers Papyrus is written in Egypt.
- 🕑 c460 BC- c370 BC – Hippocrates lived, he was a Greek physician and is known as the father of western medicine.
- 🕑 200 – Galen describes the workings of the human body, which remains unchallenged until 1500.

Sketch Assignment Week 26

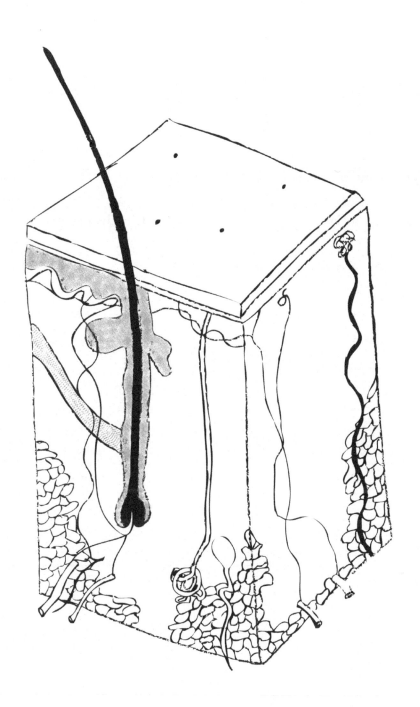

Experiment: Which is stronger, regular hair or bleached hair?

Introduction

Most mammals are covered with hair and humans are no exception. We have hair growing everywhere except our lips, the palms of our hands and the soles of our feet. Hair is made up of dead cells that are filled with a protein called keratin. In this experiment, we are going to measure the strength of unbleached and bleached hair.

Hypothesis

Circle which you think is stronger.

Unbleached Hair   Bleached Hair

Materials

_____   _____

_____   _____

_____   _____

_____   _____

_____   _____

Procedure

_____

_____

_____

_____

_____

_____

_____

_____

_____

_____

Observations

My unbleached strand of hair held _____ pennies.

My bleached strand of hair held _____ pennies.

_____

_____

_____

_____

Conclusion

_____

_____

_____

_____

_____

_____

Written Assignment Week 26

Discussion Questions

1. What are the two main layers of skin?
2. What is the job of skin?
3. Where can hair be found on the human body?
4. How does hair grow?
5. What are the three parts of a nail?

_____

_____

_____

_____

_____

_____

_____

_____

_____

_____

_____

_____

_____

_____

_____

_____

_____

_____

_____

_____

_____

Written Assignment Week 26

Student Assignment Sheet Week 27
Skeletal System

Experiment: Which bone breaks easier?

Materials

- ✓ 3 Chicken thigh bones
- ✓ Vinegar
- ✓ Glass
- ✓ Hammer
- ✓ 3 Plastic bags

Procedure (***Note** – Step 2 of this experiment takes 24 hours.*)

1. Read the introduction to this experiment and circle which bone you think will break the easiest.
2. Place one of the bones in a glass and cover it with vinegar. Let it sit for 24 hours.
3. The next day, remove and rinse the bone thoroughly. Place it in a plastic bag and label it bone #1.
4. Place another one of the bones in the oven and bake it for thirty minutes at 450°F. Remove it and let it cool. Then place it in a plastic bag and label it bone #2.
5. Place the third bone in a plastic bag and label it bone #3.
6. Bend each bone to see if they break. Then, use a hammer to hit each bone once. Be very careful not to hit yourself!
7. Examine each of the bones, record the damage and complete the experiment sheet.

Vocabulary & Memory Work

- ☐ Vocabulary: cartilage, axial skeleton, appendicular skeleton
- ☐ Memory Work—Continue to work on the Body Systems. This week add:
  - ✓ **Skeletal System** – It supports the body, protects organs and permits movement.
- ☐ Memory Work—Work on the Major Bones and their Locations.

1. Cranium	5. Humerus	9. Sternum	13. Fibula
2. Mandible	6. Radius	10. Pelvis	14. Tibia
3. Clavicle	7. Ulna	11. Femur	
4. Scapula	8. Rib cage	12. Patella	

Sketch: Major Bones of the Body

- 🖾 Label all of the bones listed in the memory work above.

Writing

- ✍ Reading Assignment: *Kingfisher Science Encyclopedia* pp. 102-103 The Skeleton
- ✍ Additional Research Reading
  - 📖 The Skeleton: *USE* pp. 346-347, *DK HB* pp. 16-17, *UIDS* pp. 278-279
  - 📖 Bones & Joints: *DK HB* pp. 18-21, *UIDS* pp. 280-281

Dates to Enter

- ☐ 1510-1580 – Giovanni Ingrassis lived; he is credited as the founder of osteology, which is the study of bones.
- ☐ 1691 – English Dr. Clopton Haves describes the structure of bones.
- ☐ 1895 – German physicist Wilhelm Roentgen discover x-rays.

Sketch Assignment Week 27

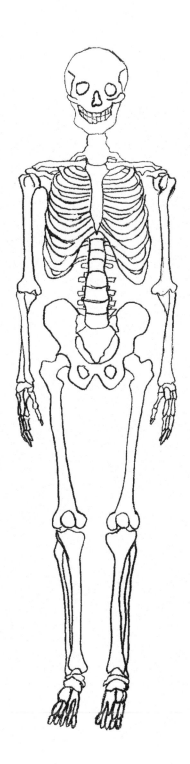

Experiment: Which bone breaks easier?

Introduction

The job of the skeletal system is to shape and support the body. Bones are the foundation of the skeletal system. Bones are made living cells that have been hardened with calcium, called osteocytes. This hardening gives them their strength. In this experiment, you will examine the strength of bones.

Hypothesis

Circle which you will break easier:

Bone soaked in vinegar   Baked Bone   Regular Bone

Materials

_____   _____

_____   _____

_____   _____

_____   _____

_____   _____

Procedure

_____

_____

_____

_____

_____

_____

_____

_____

_____

_____

Observations

Bone	Damage Received
#1-soaked in vinegar	
#2-baked in oven	
#3-regular bone	

Conclusion

_____

_____

_____

_____

_____

_____

Written Assignment Week 27

Discussion Questions

1. What are the functions of the skeletal system?
2. Name the four types of bones and give an example of each.
3. How many bones does the average adult skeleton have?
4. What do the bones of the skull do?
5. What bones make up the axial skeleton?
6. What bones make up the appendicular skeleton?

_____

_____

_____

_____

_____

_____

_____

_____

_____

_____

_____

_____

_____

_____

_____

_____

_____

_____

_____

_____

_____

Written Assignment Week 27

Student Assignment Sheet Week 28
Muscular System

Experiment: Do muscles have memory?

Materials
- ✓ Yourself
- ✓ Door frame

Procedure
1. Read the introduction to this experiment and answer the question.
2. Stand in a door frame, push the outside of your hand against the door frame for at least thirty seconds.
3. Step out of the doorway quickly and see what happens. (*Be sure not to intentionally move your arms when you step out.*)
4. Write down what happened and complete the experiment sheet.

Vocabulary & Memory Work
- ☐ Vocabulary: myofibrils, involuntary muscle, voluntary muscle
- ☐ Memory Work—Continue to work on the Body Systems. This week add:
  - ✓ **Muscular System** – It moves the body and helps support it.
- ☐ Memory Work—Continue to work on the major bones and their locations.

Sketch: Muscle Fiber & 3 Types of Muscles
- 🖾 For the muscle fiber, label the following – sheath, nucleus, myofibrils, filaments containing the actin and myosin
- 🖾 For the 3 types of muscles (at bottom of sketch sheet), label the following – skeletal muscle, smooth muscle, cardiac muscle

Writing
- ﹏ Reading Assignment: *Kingfisher Science Encyclopedia* pp. 106-107 Muscles & Movement
- ﹏ Additional Research Reading
  - 📖 Muscles: *USE* pp. 348-349, *DK HB* pp. 22-23, *UIDS* pp. 282-283
  - 📖 The Moving Body: *DK HB* pp. 24-25

Dates to Enter
- 🕐 1637-1690 AD – Dutch Jan Swammerdam lived; he found that muscles change shape and not volume during contraction.
- 🕐 1638-1686 AD – Danish scientist Neils Stensen lived; he studied muscles under a microscope and discovered what causes muscle contraction.
- 🕐 1668-1707 AD – Giorgio Baglivi (Italian) lived; he was the first to see that skeletal muscles were different from those in the organs.

Sketch Assignment Week 28

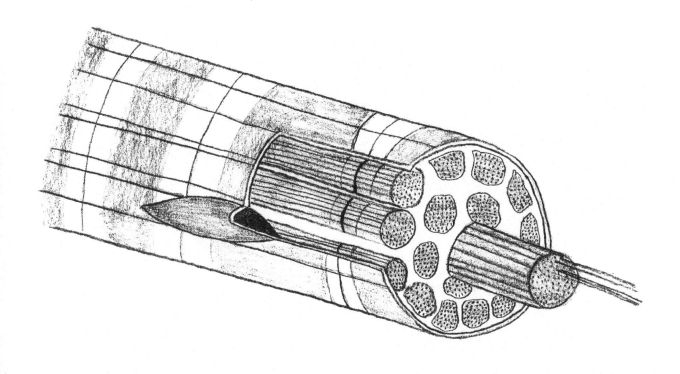

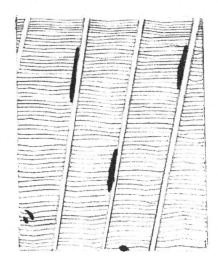
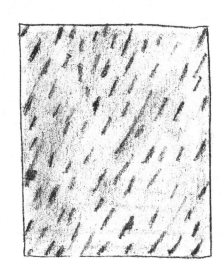
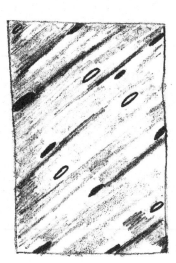

Experiment: Do muscles have memory?

Introduction

Muscles give the human body the ability to move. They are made up of many strands called myofibrils. Inside each of the myofibrils is filaments of actin and myosin. When the muscle receives the signal to contract from the brain, the two filaments slide past each other causing the muscle to shrink and contract. In this experiment, you are going to test to see if your muscles have memory.

Hypothesis

Do muscles have memory?　　　Yes　　　　　　No

Materials

_____　　　　_____

_____　　　　_____

_____　　　　_____

_____　　　　_____

_____　　　　_____

Procedure

_____

_____

_____

_____

_____

_____

_____

_____

_____

_____

_____

_____

Observations and Results

_____

_____

_____

_____

_____

_____

_____

_____

Conclusion

_____

_____

_____

_____

_____

_____

_____

_____

Written Assignment Week 28

Discussion Questions

1. What are the three types of muscles and give an example of what they help the body do?
2. How do skeletal muscles contract?
3. Why do skeletal muscles work in pairs?
4. Where is smooth muscle found? How does it contract?
5. Where is cardiac muscle found? How does it contract?

_____

_____

_____

_____

_____

_____

_____

_____

_____

_____

_____

_____

_____

_____

_____

_____

_____

_____

_____

_____

_____

_____

_____

Written Assignment Week 28

Student Assignment Sheet Week 29
Nervous System

Experiment: How many centimeters (cm) will it take for me to catch a yardstick?

Materials

✓ Yardstick

✓ Partner

Procedure

1. Read the introduction to this experiment and answer the question.
2. Sit on a chair with your partner holding the yardstick just above your hand. Hold your hand in an O-shape ready to grab the yardstick.
3. Have your partner drop the yardstick without telling you when they are going to do so. Try to grab the yardstick as soon as possible and record where you grab it.
4. Repeat this three times, find the average of the three lengths and record on the experiment sheet.
5. Complete the experiment sheet.

Vocabulary & Memory Work

☐ Vocabulary: neuron, sensory neuron, motor neuron

☐ Memory Work—Continue to work on the Body Systems. This week add:

✓ **Nervous System** – It controls the body, allows a person to think and feel.

☐ Memory Work—Continue to work on the major bones and their locations.

Sketch: Nerve cells

▨ Label the following – cell body, nucleus, axon, sheath, dendrite, axon terminal

Writing

↝ Reading Assignment: *Kingfisher Science Encyclopedia* pp. 108-109 The Brain & Nervous System

↝ Additional Research Reading

📖 Nervous system: *USE* pp. 364-365, *DK HB* pp. 26-27, *UIDS* pp. 306-309

📖 The Brain: *USE* pp. 366-367, *DK HB* pp. 28-31

Dates to Enter

🕐 1664 – English Dr. Thomas Willis describes the blood supply to the brain.

🕐 1811 – Scottish anatomist Charles Bell shows that nerves are bundles of nerve cells.

🕐 1825-1893 – Jen Martin Charest lived; he is known as the founder of neurology, the study of the nervous system.

🕐 1837 – Czech biologist Johannes Purkinje observes neurons in the cerebellum in the brain.

🕐 1861 – Paul Pierre Broca discovers the area of the brain that controls speech.

Sketch Assignment Week 29

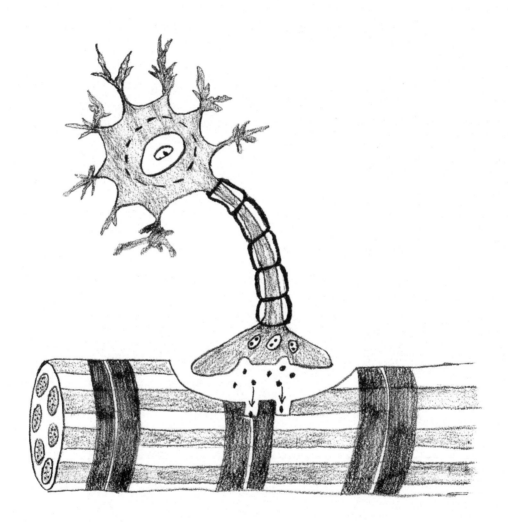

Experiment: How many centimeters (cm) will it take for me to catch a yardstick?

Introduction

Our body is controlled by a complex network of nerves that are managed by the brain. Some of these movements require us to think about them before we act while some happen automatically. The time it takes for us to think an action till when we actually do the action is called reaction time. In this experiment, you will be measuring your reaction time.

Hypothesis

I think it will take me _____ cm to catch the yardstick.

Materials

_____   _____

_____   _____

_____   _____

_____   _____

Procedure

Observations and Results

	Distance to Grab in cm
Attempt #1	
Attempt #2	
Attempt #3	
Average*	

*(add up all 3 attempts and divide that number by 3 to get your average distance)

Conclusion

_____

_____

_____

_____

_____

_____

_____

_____

_____

Optional Results from the Take it Further Activity

	Distance to Grab in cm when told by partner	Distance to Grab in cm when I told partner to drop
Attempt #1		
Attempt #2		
Attempt #3		
Average*		

Written Assignment Week 29

Discussion Questions

1. What is a neuron and what does it do?
2. What is the job of the spinal cord?
3. What controls the entire nervous system? What does each of these regions do?
4. What are the two main parts of the nervous system?

_____
_____
_____
_____
_____
_____
_____
_____
_____
_____
_____
_____
_____
_____
_____
_____
_____
_____
_____
_____
_____

Written Assignment Week 29

Student Assignment Sheet Week 30
Endocrine System (Hormones)

Experiment: Does adrenaline affect my breathing and heart rate?

Materials
- ✓ Yourself
- ✓ Watch
- ✓ Partner

Procedure

1. Read the introduction to this experiment and answer the question.
2. Sit calmly in a chair, find your pulse and count the beats for thirty seconds while your partner counts how many breaths you take in thirty seconds. Multiply both numbers by two to determine your pulse and breathing rate.
3. Now get your adrenaline going, you can do this by having your partner scare you. Then recheck your pulse and breathing.
4. Let them return to normal by resting for five minutes if your pulse and breathing increased. Then laugh hard with your partner for two minutes and recheck your pulse and breathing.
5. Write down what happened and complete the experiment sheet.

Vocabulary & Memory Work
- ☐ Vocabulary: hormone, endocrine gland
- ☐ Memory Work—Continue to work on the Body Systems. This week add:
  - ✓ **Endocrine System** – It releases hormones that control many of the body's processes.
- ☐ Memory Work—Continue to work on the major bones and their locations.

Sketch: Anatomy of the Endocrine System
- ▣ Label the following – pineal gland, pituitary gland, thyroid gland, parathyroid gland, thymus gland, adrenal gland, pancreas, ovary(female), testes(male)

Writing
- ✍ Reading Assignment: *Kingfisher Science Encyclopedia* pp. 118-119 Hormones
- ✍ Additional Research Reading
  - 📖 Hormones: *USE* pg. 363, *DK HB* pp. 40-41, *UIDS* pp. 336-337

Dates to Enter
- 🕐 1905 – British scientist Ernest Starling devises the term hormone to describe the body's chemical messengers.
- 🕐 1922 – Frederick Banting and Charles Best successfully extract insulin.

Sketch Assignment Week 30

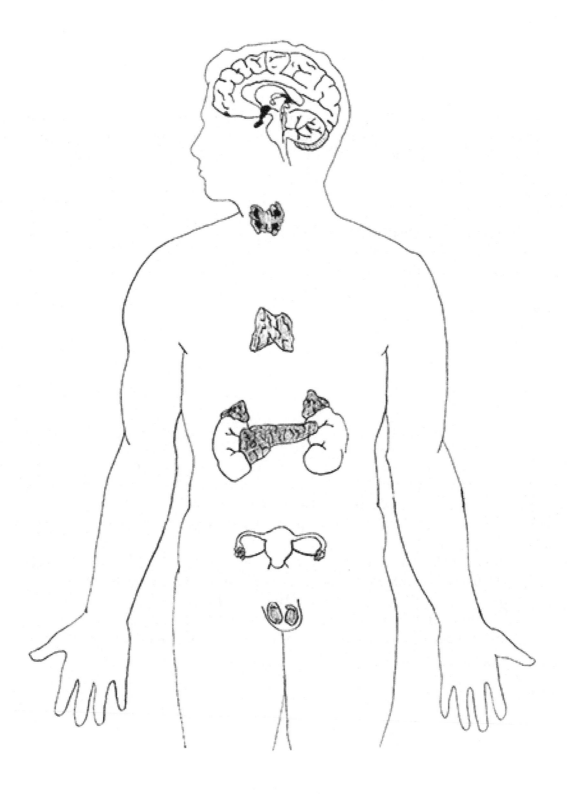

Experiment: Does adrenaline affect my breathing and heart rate?

Introduction

Hormones are the body's chemical messengers. The endocrine system produces and releases hormones that control and coordinate the functions of the body. Hormones alter the activities of the body by speeding up or slowing down the various processes. Adrenaline is a hormone that is released by the adrenal gland and it is associated with the fight or flight response. In this experiment,, you are going to test your body's response to the presence of adrenaline.

Hypothesis

Does adrenaline affect my breathing?   Yes   No

Does adrenaline affect my heart rate?   Yes   No

Materials

_____   _____

_____   _____

_____   _____

_____   _____

_____   _____

How to check your pulse

ALWAYS use two fingers to check your pulse. **NEVER** use your thumb as one of those fingers! You can find your:

✓ **Radial pulse** – The radial pulse is just inside of the wrist. Place your fingers at the base of your thumb on the inside your wrist and count the beats.

OR

✓ **Carotid pulse** – The carotid pulse is found on the side of the neck. Place your two fingers between your windpipe and the large neck muscle, press lightly until you feel the beating.

Procedure

_____

_____

_____

_____

_____

_____

_____

_____

_____

Observations and Results

	Pulse (beats per minute)	Breaths per minute
At Rest		
After being scared		
After laughing		

Conclusion

_____

_____

_____

_____

_____

_____

Written Assignment Week 30

Discussion Questions

1. What does the endocrine system do?
2. How do hormones alter the activities of a cell?
3. What organ links the nervous system and the endocrine system?
4. How are hormones regulated?
5. What does the pituitary gland do?
6. What hormones does the adrenal gland release? What do they regulate?
7. What two hormones does the pancreas release? What do they regulate?

_____

_____

_____

_____

_____

_____

_____

_____

_____

_____

_____

_____

_____

_____

_____

_____

_____

_____

_____

_____

Written Assignment Week 30

Biology Unit 6: The Human Body, part 2
Vocabulary Sheet

Define the following terms as they are assigned on the Student Assignment Sheet.

1. Arteries – _____

   _____

   _____

2. Veins – _____

   _____

   _____

3. Hemoglobin – _____

   _____

   _____

4. Inhalation – _____

   _____

   _____

5. Exhalation – _____

   _____

   _____

6. Digestive enzyme – _____

   _____

   _____

7. Villi – _____

   _____

   _____

8. Excretion – _____

   _____

   _____

9. Bacteria – _____

_____

_____

10. DNA – _____

_____

_____

11. Virus – _____

_____

_____

12. Lymphocyte – _____

_____

_____

13. Immunization – _____

_____

_____

Student Assignment Sheet Week 31
Circulatory System

Experiment: Can blood separate into solids and liquid?

Materials
- ✓ Water, Cornstarch, Flour, Sugar
- ✓ Red food coloring, Glass

Procedure
1. Read the introduction and then answer the question.
2. Mix 2 cups (475 mL) cold water (plasma), ¼ cup (60 mL) flour (red blood cells), ¼ cup (60 mL) cornstarch (platelets), ¼ cup (60 mL) sugar (white blood cells) and several drops of red food coloring in a glass. Stir well, so that all of the ingredients are suspended in the water.
3. Set the glass aside and let it sit undisturbed for thirty min.
4. Observe what has happened after thirty minutes and complete the experiment sheet.

Vocabulary & Memory Work

- ☐ Vocabulary: arteries, veins, hemoglobin
- ☐ Memory Work—Continue to work on the Body Systems. This week add:
  - ✓ **Circulatory System** – It carries materials to and from cells throughout the body.
- ☐ Memory Work—Begin to work on the Components of Blood.
  - ✓ **Red blood cells** – They have hemoglobin to carry oxygen.
  - ✓ **White blood cells** – They help the body fight disease.
  - ✓ **Platelets** – They are tiny fragments of cells that help to stop bleeding.
  - ✓ **Plasma** – The liquid portion of blood.

Sketch: Anatomy of the Heart & 3 Types of Blood Cells

- ▦ Label the following on the heart – superior vena cava, pulmonary veins, right atrium, inferior vena cava, right ventricle, coronary artery, left ventricle, left atrium, pulmonary artery, aorta
- ▦ Label the 3 types of blood cell on the bottom of the page (red cells, white cells, platelets)

Writing

- ∾ Reading Assignment: *Kingfisher Science Encyclopedia* pp. 120-122 Heart and Circulation, Blood
- ∾ Additional Research Reading
  - ▭ The Circulatory System: *USE* pp. 350-351, *DK HB* pp. 44-45, *UIDS* pp. 288-289
  - ▭ The Heart: *DK HB* pp. 42-43, *UIDS* pp. 290-291
  - ▭ Blood: *DK HB* pp. 46-47

Dates to Enter

- 🕐 1280 – Syrian doctor Ibn an Nafis shows that blood circulates the body.
- 🕐 1628 – English doctor William Harvey was the first person to show blood circulates in one direction.
- 🕐 1816 – French doctor Rene Laennec invents the stethoscope for listening to the heartbeat.
- 🕐 1901 – Karl Lantsteiner identifies the blood groups.
- 🕐 1952 – Paul Zoll designs the first cardiac pacemaker.
- 🕐 1987 – Dr. Robert Jarvik invents the first artificial heart.

Sketch Assignment Week 31

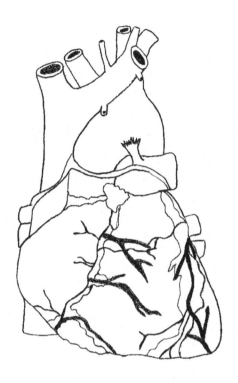

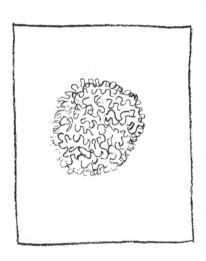
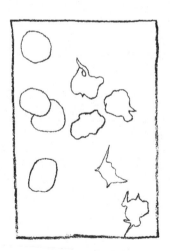

Experiment: Can blood separate into solids and liquid?

Introduction

Blood is a key component of the circulatory system. It acts as the body's delivery and removal system. Blood also helps to defend the body against viruses and bacteria. Blood has four main components, plasma, red blood cells, white blood cells and platelets. All of these components are in suspension as they are constantly pumped around the body. A suspension is when the particles of a substance are mixed with a fluid, but they remain undissolved. In this experiment, we are going to show if it is possible to separate the solids in blood from the liquid portion of blood (plasma). Your plasma will be cold water and your solids will be flour for red blood cells, sugar for white blood cells and cornstarch for platelets.

Hypothesis

Can blood separate into solids and liquid?   Yes   No

Materials

_____   _____

_____   _____

_____   _____

_____   _____

_____   _____

Procedure

_____

_____

_____

_____

_____

_____

_____

_____

_____

Observations

My suspension just after making it	My suspension after it sat for 30 minutes

_____
_____
_____
_____

Conclusion

_____
_____
_____
_____
_____
_____

Written Assignment Week 31

Discussion Questions

1. What is the job of the circulatory system?
2. What are the three main types of blood vessels?
3. What is the job of the heart?
4. Why does the heart need its own blood supply?
5. Describe the flow of blood beginning with oxygen poor blood entering the right atrium.
6. What is the job of blood?

Written Assignment Week 31

220

Student Assignment Sheet Week 32
Respiratory System

Experiment: Does increased exercise cause my breathing to increase?
> Materials
>> ✓ Watch with a second hand
> Procedure
>> 1. Read the introduction and then answer the question.
>> 2. Sit down in a chair. Count how many times you breathe in thirty seconds, then multiply that number by 2 to get how many times your breathing rate.
>> 3. Next do jumping jacks for thirty seconds. Immediately after you finish, count how many breaths you take in thirty seconds and record it.
>> 4. Rest for five minutes
>> 5. Repeat Step 3, except this time do jumping jacks for two minutes. Rest for five minutes.
>> 6. Repeat Step 3 again, except this time do jumping jacks for four minutes.
>> 7. Do the calculations (multiply by two to find your breathing rate) and complete the experiment sheet.

> ☹ **CAUTION**
>
> Do not do this experiment unless you are healthy enough for exercise. Consult a physician before beginning any exercise. If you feel short of breath, STOP!

Vocabulary & Memory Work
☐ Vocabulary: inhalation, exhalation
☐ Memory Work—Continue to work on the Body Systems. This week add:
> ✓ **Respiratory System** – It delivers oxygen into the bloodstream.

☐ Memory Work—Continue to work on the Components of Blood.

Sketch: Anatomy of the Respiratory System
▨ Label the following on the heart – bronchus, bronchiole, alveoli cluster, alveolus, blood capillaries, trachea, nasal passages

Writing
↷ Reading Assignment: *Kingfisher Science Encyclopedia* pp. 124-125 Lungs and Breathing
↷ Additional Research Reading
> 📖 The Respiratory System: *USE* pp. 358-359, *DK HB* pp. 48-49, *UIDS* pp. 298-299
> 📖 Inside the Lungs: *DK HB* pp. 50-51

Dates to Enter
🕐 1640-1679 – John Mayow lived; he performed experiments with a candle and an animal to determine that something in air is needed for breathing and burning.
🕐 1729-1799 – Lazzaro Spallareani lived; he said that respiration is done by every cell in the body.
🕐 1770 – Antoine Lavoisier names the element oxygen.

Sketch Assignment Week 32

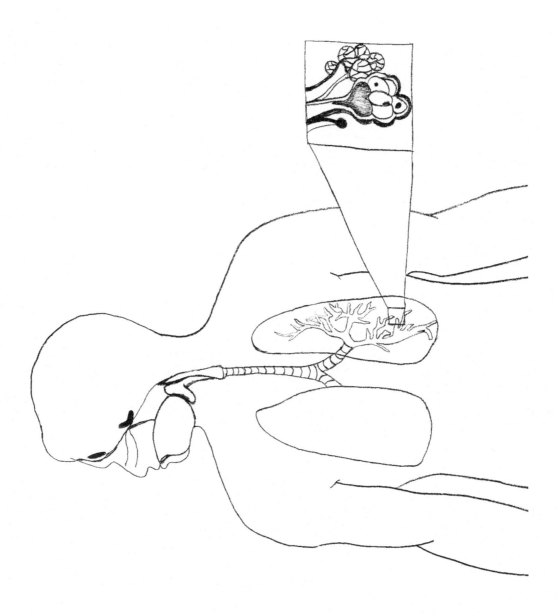

Experiment: Does increased exercise cause my breathing to increase?

Introduction

Our bodies require oxygen to survive. The oxygen we need to live comes from the air that we breathe into our lungs. One breath consists of inhalation (air is taken in) and exhalation (air is forced out). The amount of breaths you have in one minute is called your breathing rate. Your body automatically controls your breathing rate based on how much oxygen your cells need. In this experiment, you will see if the amount of time you exercise for makes a difference in how much your breathing rate increases or decreases.

Hypothesis

I think my breathing rate will (increase   /   decrease   ) the longer I exercise.

Materials

_____   _____

_____   _____

_____   _____

_____   _____

_____   _____

Procedure

_____

_____

_____

_____

_____

_____

_____

_____

_____

_____

_____

Observations and Results

Exercise time	Breaths in 30 seconds	Breathing Rate (=breaths in 30 seconds x 2)
Resting Rate		
30 seconds		
Resting Rate		
2 min		
Resting Rate		
4 min		

Conclusion

_____

_____

_____

_____

_____

_____

Written Assignment Week 32

Discussion Questions

1. What is the job of the respiratory system?
2. Why does mucus line the nasal passages and the trachea?
3. What are alveoli and where are they found?
4. What two gases are exchanged in the lungs?
5. Describe inhalation.

Written Assignment Week 32

Student Assignment Sheet Week 33
Digestive System

Experiment: How is water absorbed in the intestines?

Materials
- ✓ 3 Cups Oatmeal (cooked according to package directions and cooled)
- ✓ Water
- ✓ Bowl
- ✓ Pantyhose

Procedure

(**Note** – *Do this experiment outdoors or over a bathtub as it is quite messy!*)

1. Read the introduction and then answer the question.
2. In the bowl, mix 1 cup (240 mL) of water to the oatmeal (this is the digestive juices).
3. Spoon the mixture into the top of the pantyhose and begin gently squeezing the mixture down the leg of the pantyhose over the bowl so you can catch anything that may come out.
4. Continue squeezing until all the oatmeal mixture is in the toe of the pantyhose.
5. Measure any liquid you have collected and complete the experiment sheet.

Vocabulary & Memory Work
- ☐ Vocabulary: digestive enzyme, villi
- ☐ Memory Work—Continue to work on the Body Systems. This week add:
  - ✓ **Digestive System** – It breaks down food so that the body can use its nutrients.
- ☐ Memory Work—Continue to work on the Components of Blood.

Sketch: Anatomy of the Digestive System
- ▦ Label the following – mouth, esophagus, liver, stomach, small intestine, colon, cecum, rectum, anus, appendix

Writing
- ✎ Reading Assignment: *Kingfisher Science Encyclopedia* pp. 128-129 Digestion
- ✎ Additional Research Reading
  - 📖 Digestion: *USE* pp. 354-355, *DK HB* pp. 54-55, *UIDS* pp. 294-295
  - 📖 Teeth: *USE* pp. 352-353, *KSE* pg. 127

Dates to Enter
- 🕐 1833 – William Beaumont publishes the results of his experiments into the mechanism of digestion.
- 🕐 1848 – Claude Bernard showed that main part of digestions occurs in the small intestine and how the kidney works.
- 🕐 1871 – Wilhelm Kuhne invents the term "enzyme" to describe substances that accelerate chemical reactions.

Sketch Assignment Week 33

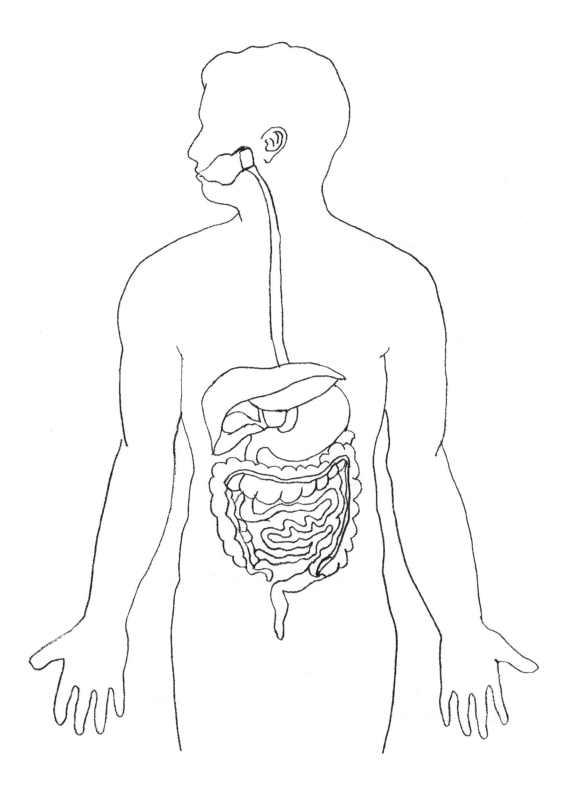

Experiment: How is water absorbed in the intestine?

Introduction

Digestion occurs in four stages. The first stage is ingestion; food is taken into your mouth, chewed and swallowed. The second stage is digestion; food is broken down by muscular crushing and enzymes in the stomach. The third stage is absorption; nutrients from the food are moved into the blood in the intestines. The final stage is egestion; waste ejected out of the body. In this experiment, you will look closer at the third stage of digestion.

Hypothesis

I think water is absorbed in the intestine by _____

_____

_____

Materials

_____   _____

_____   _____

_____   _____

_____   _____

_____   _____

Procedure

_____

_____

_____

_____

_____

_____

_____

_____

_____

_____

Observations and Results

I was able to exact out _____ cups of water from my pantyhose intestines.

_____

_____

_____

_____

_____

_____

_____

Conclusion

_____

_____

_____

_____

_____

_____

_____

_____

Written Assignment Week 33

Discussion Questions

1. What is the job of the digestive system?
2. Name the four stages of digestion and tell what happens in them.
3. Name the three functions of the stomach.
4. What are the three sections of the small intestine and what happens in them?
5. What is the main job of the large intestine?

Written Assignment Week 33

Student Assignment Sheet Week 34
Urinary System

Experiment: Do kidneys filter out liquids or solids?

Materials

- ✓ 2 cups
- ✓ 1 coffee filter
- ✓ Food coloring
- ✓ 3 Tablespoons flour
- ✓ Water
- ✓ Rubber band

Procedure

1. Read the introduction and then answer the question.
2. Attach the coffee filter to one of the cups using the rubber band, leaving a little pocket in which to pour liquid.
3. In the other cup, mix 3 TBSP (22 g) flour with ½ cup (120 mL) of water and four drops of food coloring. Stir well and slowly pour into the coffee filter pocket in the other glass.
4. Let the cup sit undisturbed for twenty minutes.
5. After twenty minutes, observe what has happened and complete the experiment sheet.

Vocabulary & Memory Work

- ☐ Vocabulary: excretion
- ☐ Memory Work—Continue to work on the Body Systems. This week add:
  - ✓ **Urinary System** – It removes waste materials.
- ☐ Memory Work—Continue to work on the Components of Blood.

Sketch: Anatomy of the Urinary System

- ▣ Label the following – aorta, inferior vena cava, right kidney, left kidney, ureter, bladder, urethra, adrenal glands

Writing

- ᨒ Reading Assignment: *Kingfisher Science Encyclopedia* pg. 131 Waste Disposal
- ᨒ Additional Research Reading
  - ▭ Balancing Act: *USE* pg. 362
  - ▭ The Urinary system: *UIDS* pp. 300-302
  - ▭ Waste Disposal: *DK HB* pp. 56-57

Dates to Enter

- ☉ 384-322 BC – Aristotle writes several books, one of which describes the urinary system and how it works.
- ☉ 1842 – British surgeon William Bowen describes the microscopic structure and working of the kidney.
- ☉ 2006 – A urinary bladder is grown in the lab from a patient's own cells and is successfully transplanted into the patient.

Sketch Assignment Week 34

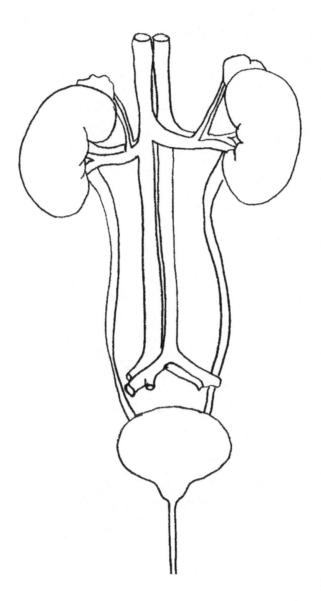

Experiment: Do kidneys filter out liquids or solids?

Introduction

The job of the urinary system is to remove waste materials from the blood and to pass those materials out of the body. It consists of the kidneys, the ureter, the bladder and the urethra. The job of the kidneys is to filter the blood to remove the waste materials. In this experiment, we are going to examine how the kidneys filter.

Hypothesis

I think kidneys filter out (liquids   solids   ).

Materials

_____   _____

_____   _____

_____   _____

_____   _____

_____   _____

Procedure

_____

_____

_____

_____

_____

_____

_____

_____

_____

_____

_____

_____

_____

Observations and Results

At the beginning	After 20 minutes of filtering

_____

_____

_____

_____

_____

_____

Conclusion

_____

_____

_____

_____

_____

_____

_____

_____

Written Assignment Week 34

Discussion Questions

1. What are the major excretory organs and what do they excrete?
2. What two things do kidneys remove from the blood?
3. What does each nephron contain?
4. How many times a day do the kidneys process the whole body's blood supply?
5. Name the three layers of the kidney and tell what each one does.

Written Assignment Week 34

Student Assignment Sheet Week 35
Immune System

Experiment: How do viruses and bacteria spread?

Materials
- ✓ Several friends
- ✓ Several different colors of glitter, one for each person

Procedure
1. Read the introduction and then answer the question.
2. Choose a color of glitter and rub some on your hands, then sneeze. Observe what has happened to the glitter and record that on the experiment sheet.
3. Have your friends choose a different color of glitter each and rub some of their glitter on their hands. Then, go around shaking each other's hands.
4. Observe what has happened, record it and complete the experiment sheet.

Vocabulary & Memory Work
- ☐ Vocabulary: bacteria, DNA, virus, lymphocyte, immunization
- ☐ Memory Work – Continue to work on the Body Systems. This week add:
  - ✓ **Immune System** – It defends the body against disease.
- ☐ Memory Work—Continue to work on the Components of Blood.

Sketch: How a Virus Reproduces
- ▨ Label the following – 1-attack, 2-invasion, 3-overthrow, 4-multiplication, 5-dispersal

Writing
- ✍ Reading Assignment: *Kingfisher Science Encyclopedia* pg. 136 Bacteria and Viruses, pg. 137 Immune System
- ✍ Additional Research Reading
  - 📖 Fighting Disease: *USE* pp. 386-387, *DK HB* pg. 45
  - 📖 The Lymphatic System: *UIDS* pp. 292-293

Dates to Enter
- 🕐 1749-1823 – Edward Jenner lived, he performed the first vaccine.
- 🕐 2002 – Gene therapy is used to treat boys suffering from an inherited immunodeficiency disease which leaves the body unable to fight against infection.

Sketch Assignment Week 35

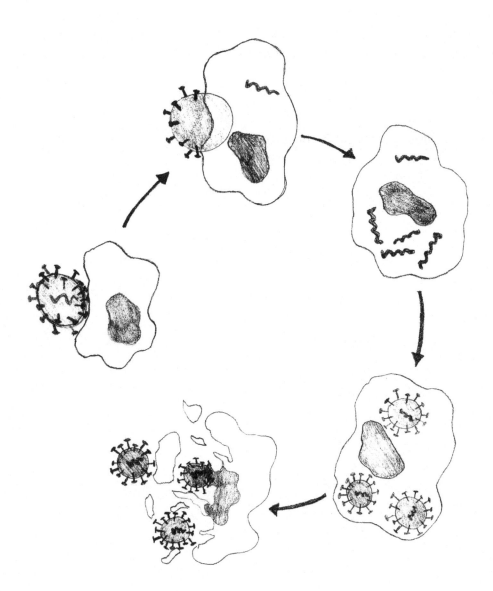

Experiment: How do viruses and bacteria spread?

Introduction

The job of the immune system is to defend the body against attacking viruses and bacteria. The body's first line of defense is the skin, but there are gaps in that, such as the mouth, nose and eyes. Viruses and bacteria on the hands can enter the body when the hand touches one of these gaps. In this experiment, we are going to look at how viruses and bacteria can be passed from person to person and from hand to hand.

Hypothesis

I think that viruses and bacteria spread by _____

_____

_____

Materials

_____   _____

_____   _____

_____   _____

_____   _____

_____   _____

Procedure

_____

_____

_____

_____

_____

_____

_____

_____

_____

Observations and Results

_____

_____

_____

_____

_____

_____

Conclusion

_____

_____

_____

_____

_____

_____

Written Assignment Week 35

Discussion Questions

Bacteria and Viruses, pg. 136

1. What are disease causing bacteria called?
2. How do bacteria invade the body?
3. How does a virus cause disease?

Immune System, pg. 137

1. What is the job of the immune system?
2. What are the three ways that the body defends itself?
3. Briefly describe the immune response.

Written Assignment Week 35

Appendix

Memory Work by Unit

Unit 1: Biological Building Blocks

Divisions of Life
1. Kingdom
2. Phylum
3. Class
4. Order
5. Family
6. Genus
7. Species

(*The following mnemonic can help you as you work on memorizing these:*
King Phillip Can Only Find his Green Shoes)

Five Kingdoms and their Basic Characteristics
1. **Monerans** – Microscopic organisms that have a simple, single cell. (*Example: Bacteria*)
2. **Protists** – A variety of complex, but mainly single-celled organisms. (*Example: Algae*)
3. **Fungi** – Organisms that absorb food and reproduce by making spores. (*Example: Molds*)
4. **Plants** – Living things that have many cells and most carry out photosynthesis. (*Example: Trees*)
5. **Animals** – Organisms made up of many cells and live by eating food. (*Example: Humans*)

Unit 2: Plants

Photosynthesis Equation
Carbon Dioxide + Water + Energy from the Sun $\longrightarrow$ Carbohydrates + Oxygen

Parts of a Flowering Plant
1. **Root** – It helps to anchor the plant and absorb nutrients.
2. **Stem** – It holds the plant up and serves as the transport system for the plant.
3. **Leaf** – It absorbs sunlight and produces energy for the plant through chlorophyll.
4. **Flower** – It is the reproductive part of the plant.
5. **Seed** – It contains the material necessary to grow a new plant.

Parts of a Flower
1. Petals
2. Sepals
3. Pistil
4. Stigma
5. Style
6. Ovary
7. Ovules
8. Stamen
9. Anther
10. Pollen

Biology for the Logic Stage Student Guide ~ Appendix

11. Filament

Unit 3: Invertebrates

Basic Phyla of the Animal Kingdom

1. **Annelids** – Animals that are worm-like and have segmented bodies (*Example: Earthworms*).
2. **Flatworms** (or Platyhelminthes) – Animals that are worm-like and have flat, unsegmented bodies (*Example: Flatworms*).
3. **Nematodes** – Animals that have round worm-like bodies with no segments (*Example: Roundworms*).
4. **Cnidarians** – Animals that live in water and have sack-like bodies with a single opening (*Example: Jellyfish*).
5. **Echinoderms** – Animals with spiny skin, sucker feet, and a five-rayed body (*Example: Starfish*).
6. **Mollusks** – Animals with soft-bodies; most have shells (*Example: Snails*).
7. **Porifera** – Animals that have perforated interior walls; most feed on bacteria (*Example: Sponges*).
8. **Arthropods** – Animals that have segmented bodies, jointed legs; most have a hard exoskeleton (*Example: Crabs*).
9. **Chordates** – Animals whose bodies are supported by a stiff rod called a notochord (*Example: Vertebrates*).

Unit 4: Vertebrates

Basic Classes of the Phyla Chordata

1. **Fish** – Cold-blooded animals that live in water; they are covered with scales and breathe through gills.
2. **Amphibians** – Cold-blooded animals that live on land and in the water; they have soft skin.
3. **Reptiles** – Cold-blooded animals that lay eggs; they are covered with scales.
4. **Birds** – Warm-blooded animals that lay eggs; they have feathers and wings.
5. **Mammals** – Warm-blooded animals that feed their young milk; they are covered with fur.

Unit 5: Animal Overview

Animal Defenses

1. **Playing Dead** – The prey acts as if they are already dead, which can shut off the hunting behavior in the predator.
8. **Making an Escape** – The prey makes a sudden dash for safety; this relies on the fact that the prey is fast and has sharp senses.
9. **Spines and Scales** – The prey is covered with spines or scales that make it difficult for the predator to eat it.
1. **Camouflage** – The prey disguises itself as something else, such as a leaf or twig, so that the predator cannot find it.
2. **Mimicry** – The prey mimics another more dangerous animal so that the predator will

leave it alone.

3. Chemical Weapons – The prey emits a poisonous or foul-smelling chemical to keep predators from eating it.

Unit 6: The Human Body

Body Systems

- ✓ **Integumentary System** – It covers & protects the body.
- ✓ **Skeletal System** – It supports the body, protects organs and permits movement.
- ✓ **Muscular System** – It moves the body and helps support it.
- ✓ **Nervous System** – It controls the body, allows a person to think and feel.
- ✓ **Endocrine System** – It releases hormones that control many of the body's processes.
- ✓ **Circulatory System** – It carries materials to and from cells throughout the body.
- ✓ **Respiratory System** – It delivers oxygen into the bloodstream.
- ✓ **Digestive System** – It breaks down food so that the body can use its nutrients.
- ✓ **Urinary System** – It removes waste materials.
- ✓ **Immune System** – It defends the body against disease.

Major Bones and thier Location

1. Cranium
2. Mandible
3. Clavicle
4. Scapula
5. Humerus
6. Radius
7. Ulna
8. Rib cage
9. Sternum
10. Pelvis
11. Femur
12. Patella
13. Fibula
14. Tibia

Components of Blood

- ✓ **Red blood cells** – They have hemoglobin to carry oxygen.
- ✓ **White blood cells** – They help the body fight disease.
- ✓ **Platelets** – They are tiny fragments of cells that help to stop bleeding.
- ✓ **Plasma** – The liquid portion of blood.

Major Bones of the Body

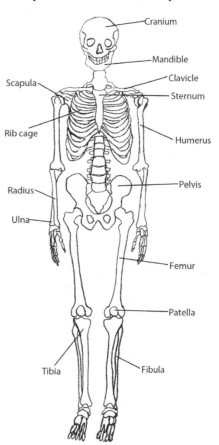

Cranium
Mandible
Clavicle
Sternum
Scapula
Rib cage
Humerus
Radius
Ulna
Pelvis
Femur
Patella
Tibia
Fibula

Activity Log

Activity	Date
What I did/saw/learned	

Activity	Date
What I did/saw/learned	

Activity	Date
What I did/saw/learned	

Activity Log

Activity	Date
What I did/saw/learned	

Activity	Date
What I did/saw/learned	

Activity	Date
What I did/saw/learned	

Activity Log

Activity	Date

What I did/saw/learned

Activity	Date

What I did/saw/learned

Activity	Date

What I did/saw/learned

Activity Log

Activity	Date

What I did/saw/learned

Activity	Date

What I did/saw/learned

Activity	Date

What I did/saw/learned

Glossary

A

- **Algae** – A simple, plant-like organism that makes its food by photosynthesis.
- **Amphibian** – A cold-blooded vertebrate that lives partly in water and partly on land.
- **Angiosperm** – A plant that reproduces by bearing flowers, fruit, and seeds.
- **Antennae** – Long sensory organs on an arthropod's head.
- **Appendicular skeleton** – The part of the skeleton made up of the bones in the shoulders, arms, pelvis and legs.
- **Arteries** – Strong vessels that carry blood away from the heart to the body.
- **Autotroph** – A living thing that makes its own food.
- **Axial skeleton** – The part of the skeleton made up of the skull, backbone and rib cage.

B

- **Bacteria** – A group of microscopic organisms without cell nuclei, many of which cause disease.
- **Bivalve** – A mollusk with a shell made of two parts or valves, such as an oyster or mussel.

C

- **Camouflage** – The way animals hide by blending in with their surroundings.
- **Carnivore** – An animal with specially shaped teeth that feeds mainly on meat.
- **Cartilage** – A tough flexible tissue that cushions joints and makes body parts such as the ears and trachea.
- **Cell** – A tiny unit of living matter, the basic building block of all life.
- **Chlorophyll** – The green chemical that gives most plants their color; it traps the sun's energy so that the plant can use it to make food.
- **Chromosomes** – Bundles of DNA that can be found in the cell's nucleus; they work together to tell an organism how to develop.
- **Classification** – A way of identifying or grouping living things.
- **Clutch** – A set of eggs.
- **Coral** – A small sea animal that catches food with stinging tentacles; many live in large colonies on coral reefs.
- **Cotyledon** – A small leaf inside a seed.

- **Crop** – A pouch in a bird's digestive system where swallowed food is stored.
- **Crustacean** – An invertebrate with jointed legs and two pairs of antennae, such as a crab.
- **Cuticle** – The protective waterproof outer layer of epidermis of an animal.

D

- **Deciduous** – A plant that sheds all its leaves during some part of the year.
- **Dermis** – The thick lower layer of skin beneath the epidermis.
- **Dicot** – A flowering plant that has two cotyledons.
- **Digestive enzyme** – The fluids produced by digestive glands which contain enzymes that break down food into simpler substances.
- **DNA** – The acid found in cell nuclei that forms the genes and chromosomes of all living things.

E

- **Echinoderm** – A sea animal with an internal skeleton and a body divided into five equal parts, such as a starfish.
- **Ecology** – The study of relationships between living things and their environment.
- **Ecosystem** – A collection of living things and their environment.
- **Ectotherm** – An animal whose temperature varies with its surroundings; ectothermic animals are also called cold-blooded.
- **Endocrine gland** – A hormone-making gland.
- **Endotherm** – An animal with a constant temperature; endothermic animals are also called warm-blooded.
- **Epidermis** – The outer layer of skin.
- **Evergreen** – A plant that keeps its leaves throughout the year.
- **Excretion** – The removal of waste substances from the body.
- **Exhalation** – Breathing out.
- **Exoskeleton** – A hard, outer skeleton that surrounds an animal's body.

F

- **Fledgling** – A young bird at the time it leaves the nest.
- **Food chain** – A process whereby energy passes along a chain of living things.
- **Food web** – A collection of interconnected food chains.
- **Frond** – The leaf of a fern or palm.

- **Fungi** – Living things that absorb food from living or dead matter around them.

G

- **Genes** – A set of instructions that tell your body how to develop; found on the chromosomes.
- **Germination** – The process in which a seed begins to grow when conditions are right.
- **Gills** – An organ used to breathe underwater.
- **Gizzard** – A chamber in a animal's stomach that grinds up food.
- **Gymnosperm** – A plant that reproduces by making cones

H

- **Habitat** – The natural home of a species.
- **Hemisphere** – One of the halves of the Earth created by an imaginary division along the equator.
- **Hemoglobin** – A purple-red chemical in red blood cells that is responsible for carrying oxygen.
- **Herbivore** – An animal that eats only plant food.
- **Heterotroph** – A living thing that eats other living things.
- **Hormone** – A chemical which helps to control the level of substances in the body.
- **Hydrophyte** – A plant that is specifically adapted to live in water.

I

- **Immunization** – A medical procedure that primes the body's immune system to fight specific infections.
- **Inhalation** – Breathing in.
- **Invertebrate** – An animal without a backbone, such as an insect.
- **Involuntary muscle** – Muscles that act automatically, like the heart.

J

K

- **Kingdom** – The highest category into which living things are classified.

L

- **Larva** – A young animal that develops into an adult by a complete change in body shape.

- **Life cycle** – The pattern of changes that occurs in each generation of a species.
- **Lymphocyte** – A white blood cell that destroys germs by releasing antibodies.

M

- **Mammals** – A warm-blooded animal with hair that feeds its young on milk.
- **Mammary gland** – The milk-producing organ of a female mammal.
- **Medusa** – The umbrella-shaped, swimming stage in the life cycle of jellyfish and other cnidarians.
- **Metamorphosis** – A major change in the animal's body shape during its life cycle.
- **Migration** – A journey by an animal to a new habitat.
- **Mitosis** – The division of a cell nucleus to produce two identical cells.
- **Mollusk** – A soft bodied invertebrate that is often protected by a hard shell.
- **Monocot** – A flowering plant that has one cotyledon.
- **Monotreme** – A mammal that lays eggs, such as a duck-billed platypus.
- **Motor neuron** – Neurons which pass instructions from the central nervous system to the body.
- **Myofibrils** – Long, rod-shaped cells which make up muscle tissue.

N

- **Neuron** – A nerve cell.
- **Nutrient** – Any material that is taken in by a living thing to sustain life.
- **Nymph** – An immature insect that resembles an adult but has no wings.

O

- **Omnivore** – An animal that eats both plant and animal food.

P

- **Photosynthesis** – A process that uses light energy to make food from simple chemicals.
- **Pollination** – The transfer of pollen from the male part of a plant's flower to the female part.
- **Polyp** – A small sea animal with a hollow cylindrical body and a ring of tentacles around its mouth; one of the two stages in the life cycle of cnidarians.
- **Predator** – An animal that kills and eats other animals.
- **Prey** – An animal that is killed and eaten by other animals.

Q

R

- **Reptile** – A cold-blooded vertebrate with scaly skin.

S

- **Scales** – Small, overlapping plates that protect the skin.
- **Sensory neuron** – Neurons ending in sensitive receptors, they carry information to the central nervous system.
- **Species** – A group of living things than can breed together in the wild.
- **Spore** – A microscopic package of cells produced by a fungus or plant that can grow into a new individual.
- **Swim bladder** – A gas-filled bag that helps a fish to float in water.
- **Symbiosis** – A close ecological relationship between two different species.

T

- **Tadpole** – The immature form of a frog or toad.
- **Thorax** – The central body part of an arthropod (between the abdomen and the head).

U

V

- **Veins** – The vessels carrying blood to the heart.
- **Vertebrate** – An animal with a backbone.
- **Villi** – Tiny, finger-like projections on the lining of the small intestine which absorb digested food.
- **Virus** – Strands of DNA in a protective coat, they invade living cells and cause disease.
- **Voluntary muscle** – Muscles which can be consciously controlled, like the arm.

W

X

Y

- **Yeast** – A microscopic, single-celled fungus.

Z

Made in the USA
Columbia, SC
02 August 2021

42586197R00143